18th OpenFOAM Conference 2023

IOP Conference Series: Materials Science and Engineering
Volume 1312

Genoa, Italy
11-14 July 2023

ISBN: 979-8-3313-0268-9
ISSN: 1757-8981

Printed from e-media with permission by:

Curran Associates, Inc.
57 Morehouse Lane
Red Hook, NY 12571

Some format issues inherent in the e-media version may also appear in this print version.

Institute of Physics
Dirac House, Temple Back
Bristol BS1 6BE UK

Phone: 44 1 17 929 7481
Fax: 44 1 17 920 0979

techtracking@iop.org

Additional copies of this publication are available from:

Curran Associates, Inc.
57 Morehouse Lane
Red Hook, NY 12571 USA
Phone: 845-758-0400
Fax: 845-758-2633
Email: curran@proceedings.com
Web: www.proceedings.com

TABLE OF CONTENTS

18th OpenFOAM Workshop (18OFW)
11-14 July 2023 Genoa – Italy

The OpenFOAM Workshop is the longest-running conference on CFD with OpenFOAM® and foam-extend. Since 2006, the Workshop has been scheduled each year. Past workshops have been held at locations around the world. You can find information about past workshops on the event website https://openfoamworkshop.org/. The Workshop has been attended by OpenFOAM users and developers from all continents. The main event of the Workshop is a multi-day conference, featuring presentations and posters on industrial applications and academic research. Additionally, there is a full day dedicated to training sessions, where experienced OpenFOAM instructors provide hands-on tutorials for all levels of experience with OpenFOAM® and foam-extend. A community day offers the opportunity to meet people with similar areas of interest in "birds-of-a-feather" sessions and work-group meetings. The OpenFOAM Workshop provides a unique opportunity to connect and share experiences with other OpenFOAM users and provides a nurturing ground for open discussions and future projects.

The 18th OpenFOAM Workshop was held in Genoa, Italy, from July 11 to 14, 2023. The event took place at the Polytechnic School of the University of Genoa and was organized by the Department of Civil, Chemical and Environmental Engineering (DICCA), in collaboration with Wolf Dynamics (http://www.wolfdynamics.com/) and the OpenFOAM Workshop Committee Members. The conference was co-chaired by Prof. Jan Pralits and Prof. Joel Guerrero, both from the University of Genoa.

More information about the 18th OpenFOAM Workshop can be found at the following links:

- 18th OpenFOAM workshop website - https://bit.ly/3KzmAfN
- Full Agenda | 18 OpenFOAM Workshop - https://bit.ly/3x6V8mp
- 18th OpenFOAM Workshop | Book of unedited abstracts - https://bit.ly/4a7xqEl
- YouTube channel | Registered sessions - https://bit.ly/3RjrDEG

The 18th OpenFOAM Workshop in numbers:

- 394 attendees, of which 283 were in person and 111 online.
- 20 online grants (thank you very much to all sponsors).
- 20 community-driven training sessions.
- 2 ESI-OpenFOAM workshops.
- 5 keynote lectures: Jennifer Keenahan, Hrvoje Jasak, Joaquim Martins, Antonio Traverso, Neil Ashton.
- Gender equality (in person): 78% male, 22% female participants.

IOP Conf. Series: Materials Science and Engineering **1312** (2024) 011001 doi:10.1088/1757-899X/1312/1/011001

- Attendees by country (top five): Italy 66, Germany 48, UK 31, USA 21, France 19. Special mention to Japan and South Korea, who added 19 together.
- Presentations: 161 oral and 23 poster presentations. We received more than 200 abstracts.
- We received presentations covering 22 different topics. To name a few: multiphase 20, FSI 14, HPC 12, Biofluids 10, civil and wind engineering 10, coastal engineering and offshore applications 9, optimization and ML/AI 9, combustions, reacting flows, and sprays 10.
- 25 chairpersons: Tessa Uroic, Jennifer Keenahan, Chiara Pesci, Vishal Acharya, Joaquim Martins, Valerio D'Alessandro, James Feng, Ivan Spisso, Andre Weiner, Marco Mazzuoli, Marco Cavazzuti, Philip Cardiff, Johan Roenby, Holger Marshall, Hakan Nilsson, Marko Horvat, Massimiliano Burlando,Timofey Mukha, Miguel Nobrega, Bruno Ramoa, Riccardo Rossi, Chris Sideroff, Edoardo Alinovi, Jan Pralits, Joel Guerrero. A big thank you to all of you for keeping the event on schedule.
- 13 sponsors (https://bit.ly/47S7on8). A big thank you to all of you for making this event possible and supporting the online grants.
- For the first time, a validation challenge was organized during the workshop (https://bit.ly/3VBzmkf).

Scientific and Technical Local Organizing Committee:

Joel Guerrero. DICCA-University of Genoa and Wolfdynamics, Italy.
Jan Pralits. DICCA-University of Genoa and Wolfdynamics, Italy.
Massimiliano Burlando. DICCA-University of Genoa, Italy.
Alessio Ricci. University School for Advances Studies IUSS Pavia, Italy.
Chandan Bose. University of Edinburgh, UK.
Federico Piscaglia. Dept. of Aerospace Science and Technology (DAER), Politecnico di Milano, Italy.
Valerio D'Alessandro. Università Politecnica delle Marche, Italy.

Logistics Local Organizing Committee:

Gabriela Diaz. Start-Here S.r.l., Italy.
Anna Mascheroni, Italy.
Daniela Fiori, Italy.

Peer Review Statement

All papers published in this volume have been reviewed through processes administered by the Editors. Reviews were conducted by expert referees to the professional and scientific standards expected of a proceedings journal published by IOP Publishing.

- **Type of peer review:** Single Anonymous
- **Conference submission management system:** Morressier
- **Number of submissions received:** 17
- **Number of submissions sent for review:** 17
- **Number of submissions accepted:** 13
- **Acceptance Rate (Submissions Accepted / Submissions Received × 100):** 76.5
- **Average number of reviews per paper:** 1
- **Total number of reviewers involved:** 5
- **Contact person for queries:**
  Name: Joel Guerrero
  Email: joel.guerrero@unige.it
  Affiliation: University of Genoa, Italy.

IOP Conf. Series: Materials Science and Engineering **1312** (2024) 012001 doi:10.1088/1757-899X/1312/1/012001

On the comparison between topological and surface sensitivities for bio-fluid dynamics applications

Eric Segalerba[1], **Emanuele Gallorini**[2], **Maurizio Quadrio**[2], **Jan Pralits**[1]

[1]Dipartimento di Ingegneria Civile, Chimica ed Ambientale (DICCA), Università di Genova

[2]Dipartimento di Scienze e Tecnologie Aerospaziali, Politecnico di Milano

E-mail: `eric.segalerba@edu.unige.it`

Abstract. Nasal Breathing Difficulties (NBD) consist of pathologies which are difficult to diagnose, heal and affect a high percentage of the population; the only way to treat these diseases is, most of the times, by surgery but the success rate is not very high since many patients have no relief from it. This should be attributed to the variety of the nose's functions, which is far more complex than a simple duct to connect the environment to the lungs. CFD is emerging as an innovative approach to study these problems systematically; in this contribution the target is to use a method based on adjoint equations to get senstivity maps whose purpose is to provide surgeons with more informations on which are the regions, inside the nose, to focus on. This work focuses only on the reduction of the pressure drop across the nose, therefore a single objective function is used. Three different frameworks, with two formulations of the continuous adjoint equation method, are compared: in two cases topological sensitivities were used while in the latter the surface sensitivity was evaluated.

1. Introduction

A consistent part of the world population has some anatomic deformity of the nose and those are usually difficult to diagnose [1, 2]. Clinicians face the issue of understanding if the deformities that are detected in the medical examination are the cause of patient's symptoms and many times the answer is not straightforward.

This leads surgeons to rely on their experience, without an objective tool to choose whether to do surgery and, thus, errors are unavoidable [3]; CFD enables diagnosis of NBD and surgery planning [4].

The anatomy considered in the present study is derived from the CT scan of a 67 years old patient and to ensure a high realism of the surgery, it is executed under the guidance of an ENT surgeon.

The sensitivity analysis is performed using a method based on adjoint equations; the governing equations of the problem are the steady state, incompressible, isothermal Navier-Stokes equations and it is also made the assumption of frozen turbulence.

The numerical solution of the flow equations is calculated using the open source finite volume solver OpenFOAM. The flow is forced in the domain by setting a constant flow rate at the inlet. The inlet patch consists in a sphere placed in front of the nose and results in a good tradeoff between accuracy and computational cost [5].

Three different frameworks, with two formulations of the continuous adjoint, are compared: the formulations are "Topological Sensitivity" and "Surface Sensitivity".

2. Methods

The adjoint equation method allows for a computationally cheap way of calculating the gradient of an objective function with respect to one or more control variables; this means that an objective function must be defined. In this work a single objective function is considered and it is the pressure drop across the domain (from inlet to the outlet) that can be expressed as:

$$\begin{aligned} \mathcal{J}_{p_t} &= \int_{S_{I,O}} \left[-\left(p + \frac{1}{2} v_i^2 \right) v_j \right] n_j dS = \\ &= \int_{S_O} \left[-\left(p + \frac{1}{2} v_i^2 \right) v_j \right] n_j dS - \int_{S_I} \left[-\left(p + \frac{1}{2} v_i^2 \right) v_j \right] n_j dS \end{aligned} \tag{1}$$

Assume an optimisation problem expressed by the objective function $\mathcal{J}$ to be minimized, controlled by the variables $b_n\,(n = 1, ..., N)$ and constrained by the state equations $R_i\,(i = 1, ..., E) = 0$, with $\Psi_i\,(i = 1, ..., E)$ the adjoint variables. It is possible to write an augmented objective function (or Lagrangian) $\mathcal{L}$ as:

$$\mathcal{L} = \mathcal{J} + \int_V \Psi_i R_i dV \tag{2}$$

The Lagrangian and the objective function are formally the same; since a gradient with respect to the control variable is sought, the derivative is calculated:

$$\begin{aligned} \frac{d\mathcal{L}}{db_n} &= \frac{d\mathcal{J}}{db_n} + \frac{d}{db_n} \int_V \Psi_i R_i dV = \\ &= \frac{d\mathcal{J}}{db_n} + \int_V \Psi_i \frac{dR_i}{db_n} dV + \underbrace{\int_S \Psi_i R_i n_k \frac{dx_k}{db_n} dS}_{\text{Leibniz term}} \end{aligned} \tag{3}$$

The governing equations, in the residual form, for the two formulations are different for a term named *Brinkman penalization term* here called *top*:

$$\begin{cases} R^p &= -\dfrac{\partial v_j}{\partial x_j}, \\ R_i^v &= v_j \dfrac{\partial v_i}{\partial x_j} - \dfrac{\partial \tau_{ij}}{\partial x_j} + \dfrac{\partial p}{\partial x_i} + \underbrace{\alpha v_j}_{\text{top}}, \quad i = 1, 2, 3 \end{cases} \tag{4}$$

Term *"top"*, in Eq. 4, is controlled through the porosity α; for large values of this term the velocity tends to zero and the cells where it is large behave as a solid while when it tends to zero, the governing equations for the fluid are recovered.

The adjoint equation for the case of surface sensitivity is similar to the one for the case of topological sensitivity. The only difference is a term proportional to the porosity α, see Eq. 5, which is present only in the case of topological sensitivity. This term is denoted *adj-top* in Eq. 5, for which a complete derivation is found in [6].

$$\begin{cases} R^q &= -\dfrac{\partial u_j}{\partial x_j} + \left.\dfrac{\partial \mathcal{J}}{\partial p}\right|_V \\ R_i^u &= u_j \dfrac{\partial v_i}{\partial x_j} - \dfrac{\partial (u_i v_j)}{\partial x_j} - \dfrac{\tau_{ij}}{\partial x_j} + \dfrac{\partial p}{\partial x_i} + \underbrace{\alpha u_j}_{\text{adj-top}} + \left.\dfrac{\partial \mathcal{J}}{\partial u_j}\right|_V, \quad i = 1, 2, 3 \end{cases} \tag{5}$$

In the literature the Leibniz term is usually ignored; for the case of Topological Sensitivity it is possible to do so under the assumption of fixed boundaries, for which $dx_k/db_n = 0$. In the Surface Sensitivity the term is neglected assuming that the governing equations are satisfied along the boundaries but it is only true in the case of very fine meshes while in the other cases the error due to this assumption can be important.
The limitation of this formulation can be solved by using the Enhanced Surface Integral, or E-SI, [7]. In this formulation the Leibniz term is replaced with its volumetric counterpart (V_{term}):

$$\int_{S_W} \left(u_i R_i^v + q R^p\right) \frac{dx_k}{db_n} n_k dS = \\ = \int_V \frac{\partial}{\partial x_j} \left\{ -u_i v_j \frac{\partial v_i}{\partial x_k} - u_j \frac{\partial p}{\partial x_k} - \tau_{ij}^a \frac{\partial v_i}{\partial x_k} + u_i \frac{\partial \tau_{ij}}{\partial x_k} + \frac{\partial v_j}{\partial x_k} \right\} \frac{dx_k}{db_n} dV \tag{6}$$

in order to calculate exact second derivatives, in the governing equations along the boundaries, especially with unstructured grids is questionable. The cost of the V_{term} scales with the number of design variables because it is necessary to calculate dx_k/db_n for all the design variables. To alleviate the need of its calculation, the adjoint is enhanced by solving the adjoint to an hypotetical grid displacement PDE which is a Laplace equation, named in the paper gdPDE:

$$R_i^m = \frac{\partial^2 m_i}{\partial x_j^2} = 0 \qquad m_i = m_i\left(x_k, b_n\right) \tag{7}$$

3. Results

3.1. Surface Sensitivity

The surface sensitivity shown in Figure 1, obtained using the E-SI formulation, gives indications, expressed as gradients, on the surface regions to operate on to minimize the pressure drop across the domain.

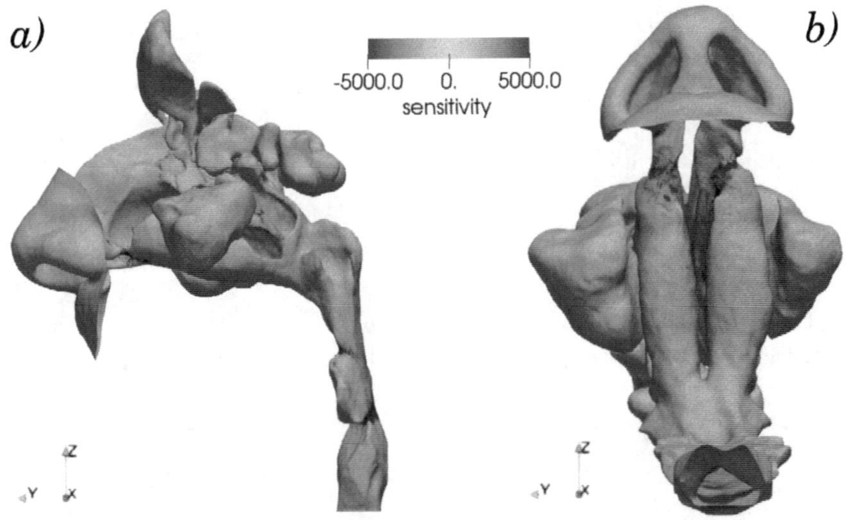

Figure 1. Surface sensitivity. The *red* stands for push out while *blue* means pull in. *a)* is the sagittal view, *b)* the bottom view.

It can be noticed that, especially from the bottom view, there are oscillations on the sensitivity which is related to the roughness of the nasal surface; to reduce pressure drop the adjoint aims to flatten the surface.

3.2. Topology Sensitivity

Figure 2 shows the results of the Topological Sensitivity evaluated in the nose. Topological optimisation

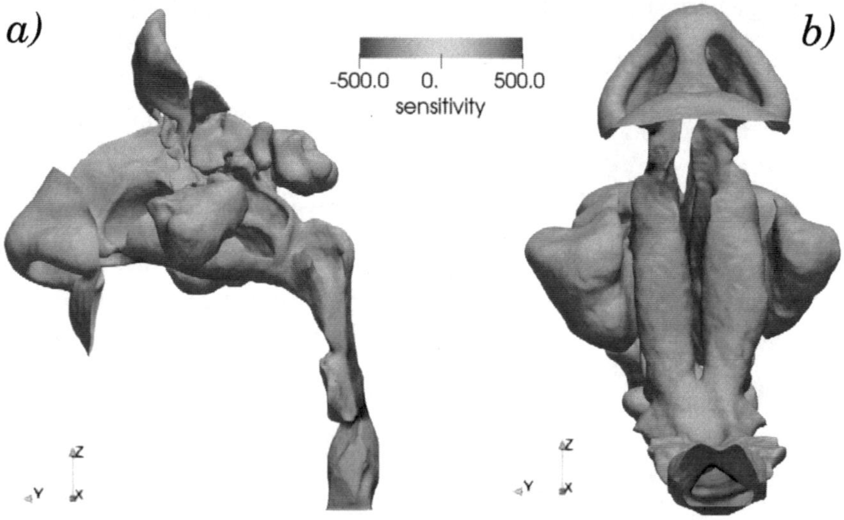

Figure 2. Topological sensitivity. The *red* means to add material while *blue* means to remove. *a)* is the sagittal view, *b)* the bottom view.

is performed by assigning a momentum loss to cells that have a negative impact on the cost function. In this case the colors represents cells which are counterproductive *blue* or productive *red*.
It can be noticed that there are less oscillations with respect to the previous case and this can be related to the fact that topological sensitivity is less influenced by the shape of the surface since the gradient is evaluated in the whole cell rather than the faces.

3.3. Topology Sensitivity with Surgery Box

The topological sensitivity can not fulfill its full capabilities without an "external" material on which to operate, so an external box called "Surgery Box" is added (Figure 3).
Using the porosity α the transition from solid to void can be controlled; while in the fluid domain this value is set to 0, in the box it is set to an arbitrary high value with the purpose of causing a large momentum loss which mimics how a solid behaves.
The coronal and sagittal planes show some interesting results: despite being trivial, in fact the minimization of the pressure loss aims to have a larger duct, it should be noticed that some topological modifications of the nose are "proposed" and these informations can not be obtained using the Shape Optimisation but are something that is typycally found in actual surgeries.

4. Conclusions

In conclusion two different formulations of the adjoint equation method are compared using the same objective function and the same boundary conditions. From these simulations the surface sensitivity appears to have oscillations related to the mesh while the topological approach is more robust.
The topological sensitivity with only the inner domain limits the capabilities of this technique which,

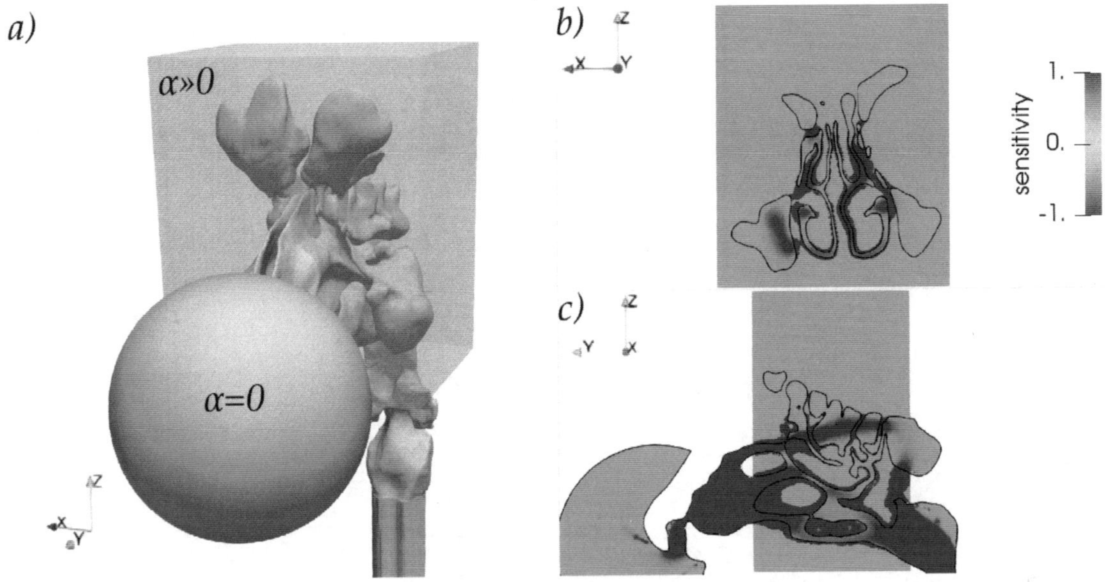

Figure 3. Topological sensitivity evaluated on a domain with an external region with high porosity. *a)* show a 3-D view of the domain: the surgery box is the transparent volume while the white represents the void. *b)* is a coronal view of the solution and *c* a para-sagittal one. As for previous Topological Sensitivity case, the *blue* stands for counter-productive and the *red* for productive.

instead, are fullfilled when an external domain is added giving interesting results which are closer to how surgeons actually operate this anatomical region.

Acknowledgments
We express our sincere gratitude to the OpenFOAM Workshop 18 organizing committee for their invaluable support.

Data Availability

Availability of data and materials Cases (including meshes), comparison data, and calculation results presented in this paper are available from the corresponding author upon reasonable request. The modified OpenFOAM version used in this work is not available for sharing.

References

[1] K. Smith, R. Orlandi, and L. Rudmik, "Cost of Adult Chronic Rhinosinusitis: A Systematic Review," *The Laryngoscope*, vol. 125, 2015.
[2] L. Rudmik, Z. M. Soler, J. C. Mace, R. J. Schlosser, and T. L. Smith, "Economic evaluation of endoscopic sinus surgery versus continued medical therapy for refractory chronic rhinosinusitis," *The Laryngoscope*, vol. 125, no. 1, pp. 25–32, 2015.
[3] C. Sundh and O. Sunnergren, "Long-term symptom relief after septoplasty," *Eur Arch Otorhinolaryngol*, vol. 272, pp. 2871–2875, Oct. 2015.
[4] P. D. Inthavong, N. Singh, and J. Sznitman, "In silico approaches to respiratory nasal flows: A review," *Journal of Biomechanics*, vol. 97, 2019.
[5] E. Segalerba, G. D. Ciacci, M. Quadrio, and J. O. Pralits, "On the comparison between pre- and post-surgery nasal anatomies via computational fluid dynamics," *Biomechanics and Modeling in Mechanobiology*, 2023.
[6] E. Gallorini, J. Hèlie, and F. Piscaglia, "An adjoint-based solver with adaptive mesh refinement for efficient design of coupled thermal-fluid systems," *Numerical Methods in Fluids*, 2023.

[7] I. Kavvadis, E. Papoutsis-Kiachagias, and K. Giannakoglou, "On the proper treatment of grid sensitivities in continuous adjoint methods for shape optimization," *Journal of Computational Physics*, 2015.

AEROSPIKE: MISSION SIMULATION AND OPTIMAL DESIGN USING OPENFOAM AND DAKOTA

Roberto Carbone, Joel Enrique Guerrero Rivas, Eric Segalerba, Jan Oscar Pralits
Dipartimento di Ingegneria Civile, Chimica ed Ambientale (DICCA), Università di Genova
E-mail: `roberto.carbone@edu.unige.it`

Abstract. Hypersonic flight presents several difficulties linked to the important viscous dissipation of kinetic energy on the surface of a vehicle which causes the need for relevant Thermal Protection Systems, reduces communication efficiency due to air ionization and creates an important pressure load due to the stagnation point on the fuselage. Those effects are enlarged by the presence of a shock wave with its strength growing in the case of a blunt body-shaped vehicle. The aim of this work is to study the effectiveness of the static device called "Aerospike" to reduce the effects induced by the hypersonic flight of a blunt body, in particular with respect to the average surface temperature, the pressure load and drag coefficient. After having studied a base case, an optimization process to define an optimal shape both for the aerospike and the aerodisk are presented.

1. Introduction

While hypersonic flight is becoming an important field for aeronautics and aerospike, the particular flow condition the aircraft is called to overcome presents several issues that have to be taken into account.

The are two main classes of hypersonic aircraft:

(i) Cruise aircraft, including missiles, rockets and drones;
(ii) Re-entry vehicles;

Those two present very different design approaches due to the particularity of the mission they have to accomplish: while, for a re-entry vehicle the speed reduction and safe landing issues are important, a cruise vehicle (in particular for drones or manned aircraft) have to take into account parameters that are similar to those considered for subsonic cruise flight. For instance drag control at the design point (related to fuel consumption and efficiency in general) which isn't a key factor for a re-entry vehicle since its main purpose is to land while guaranteeing safe conditions for the payload and/or the crew.

One of the main characteristics for an hypersonic vehicle (in particular for missiles, rockets and re-entry vehicles) is the presence of a blunt body at the stagnation point on the fuselage: this architecture is mandatory to allow the presence of a reasonable inner space to store the payload (considering the avionics/electronics or scientific equipment) and the crew (if the vehicle is manned).
For particular kind of applications, the dome-shaped forebody can be used as a static system to decrease speed in order to allow a safe landing (as done for re-entry vehicles).

The presence of this dome-shaped part induces several undesired collateral effects, for instance:

- an increase of the pressure load on the body;
- ablation of hull covering materials;
- mechanical failures;
- heat damage to the onboard electronic packages;
- communication failure due to air ionization ahead of the vehicle;

Those effects are related to the shock wave onsets near the dome that induces an important drag and temperature increase in the near-body region. This is particularly crucial during launch and cruise phases.

One of the possible solutions is to install a static device, called *Aerospike* [1], to reduce aforementioned effects. It creates a more streamlined shape reducing compressibility-related effects thanks to the insurgence of an oblique shock wave instead of a normal one. Moreover, shock effects are detached from the main body thus contributing to the protection of surface materials and payload with a weaker impact due to the onset of an oblique wake (and not a normal shock as in the blunt body case). Further, it creates a low pressure separated flow region back the head of the spike involving a pressure field onto the dome weaker than the one without the aerospike.

The head of the spike may be simply point-shaped or other kind of heads, called aerodisks [2], may be used; one of these, the cylinder-shaped aerodisk, is going to be used here to develop an optimal-designed aerospike. The chosen design is presented in figure 1 based on what was proposed by Huebner L.D. et al. [3] .

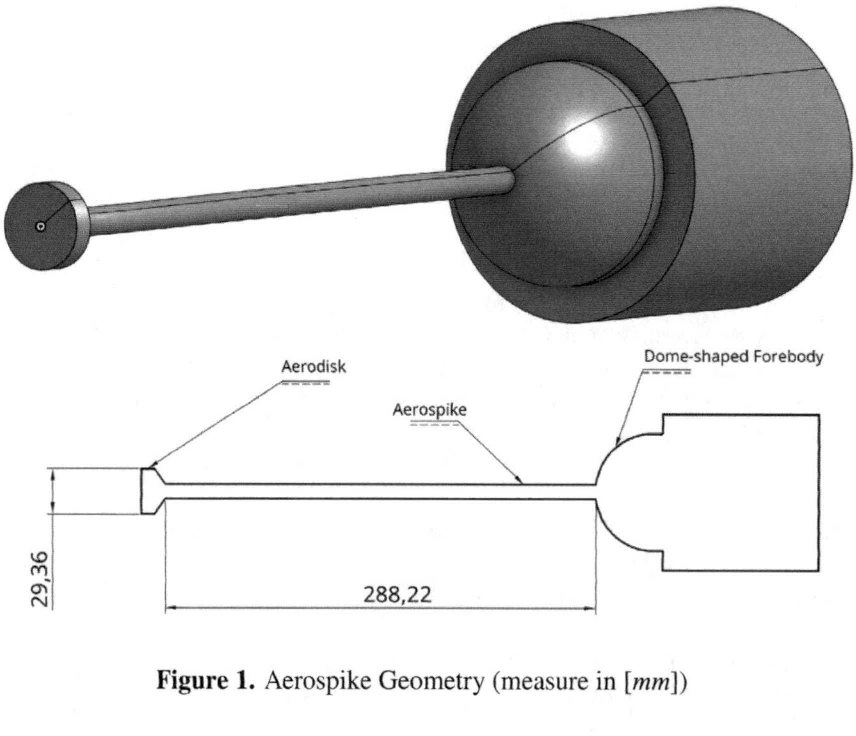

Figure 1. Aerospike Geometry (measure in [mm])

To be effective, this device has to be carefully designed. In particular, there are two main degrees of freedom that have to be calibrated:

(i) Spike length;

(ii) Aerodisk diameter;

The starting point of the process is a preliminary simulation to evaluate the installation effect onto a dome-shaped body flying at hypersonic cruise speed. The calculation is performed using OpenFOAM, the validation is done by comparing numerical results both with experimental data [3] and numerical outcomes from a validated source. The optimization process is focused on finding an optimal design for an aerospike with an cylinder-shaped aerodisk to be used for a cruise vehicle. For this case, the objective functions to be set are:

- Drag reduction;
- Static temperature control on the dome;

While the first is more related to efficiency and performances, the second regards safety, mechanical failure control, damage control, communication capacity ensuring as mentioned before. The optimisation branch is performed using Dakota.

2. Numerical model:

Hypersonic flows and, in general, supersonic conditions, are subject to important compressibility effects that causes steep density gradients in the flow field. In general, those conditions induces convergence and stability difficulties both with pressure and density based solvers. Reference values (Table 1) used

Quantity	Value	Units
Ambient pressure	1950	*Pa*
Mach number	6	*Ad*
Ambient Temperature	58	*K*

Table 1. Aerospike operating conditions

for those cases are computed starting from what described in [3] using the assumption of adiabatic flux. Non dimensional parameters used during those set of experiences are y^+, *Mach number* and *Drag coefficient* defined by:

$$y^+ = \frac{U_\tau y}{\nu}$$

$$Ma = \frac{U}{\sqrt{\gamma p \rho}}$$

$$c_d = \frac{F_D}{\frac{1}{2}\rho u^2 A}$$

where:

- U_τ is the shear velocity;
- y is the y coordinate;
- ν is the kinematic viscosity;
- U is the velocity magnitude;
- γ heat coefficient ratio;
- p is the pressure;

- ρ is the density;
- F_D is the Drag;
- A is the characteristic Area (the projection of the Dome).

The domain surrounding the body is a square-shaped, 2D fluid region in which the inlet is on the left side and the outlet (split in two parts with the same conditions) is on the right side, with respect to the orientation shown in Figure 2. The shape has been chosen to satisfy the need of having the boundaries of the domain far from the region where the shock wave onsets to avoid reflection or instabilities and diffuse the compressibility effects minimizing the interaction with the outlet.
As a consequence, the horizontal part of the device has been elongated to reach the outlet: this does not influence the solution because we are investigating the aerospike and it's effects on the Dome: due to hypersonic conditions perturbation only propagates upstream thus a change in the geometric configuration downstream doesn't have an influence on the upstream condition on the Dome or on the aerodisk where the shock wave onsets.
Ambient conditions are reported in table 1. For the temperature field, an *InletOutlet* condition was used for the outlet, the upper and the lower bound, a zero gradient was used for the walls and a fixed value for the inlet. For the pressure field, the outlet was defined as *waveTrasmissive* and the other patches were defined with a *zeroGradient*. The velocity field was initialized with a *supersonicFreestream* for the inlet, a *slip* condition on the top and bottom wall, an *inletOutlet* at the outlet and a *noSlip* condition on the walls. For the steady setup, the *waveTrasmissive* condition has been replaced with a *fixedValue* condition.
Doe to convergence issues, the steady solution was successively initialized with the converging solution of the steady solver.

The computational mesh was made using SnappyHexMesh. To simulate a 2D case, the resulting mesh was extruded to obtain one layer in the third dimension.
The resulting mesh has 2092727 cells with 10 inflection layers. In table 2 are summarized the mesh quality metrics.

Metric	Value
Number of cells	2092727
Max non-orthogonality	57.7
Average non-orthogonality	2.37
Max skewness	2.44

Table 2. Mesh quality metrics

A convergence test was made to evaluate the grid independence and the resulting mesh is visible in figure 2.
Both steady and unsteady simulation were set up. For the first, the solver used was rhoSimpleFoam and for the second one rhoPimpleFoam. While the steady situation struggles to converge to a stable solution, the unsteady one converges and the results were validated both looking at experimental results and, in particular, by comparison with results from a validated solver. The discretization schemes used were *cellLimited Gauss linear* for the gradient and *Gauss linearUpwind* for the others with default options. Both for the PIMPLE and the SIMPLE loops transonic and consistent option were activated. Relaxation factors were one of the options calibrated to archive a good converging solution: for the unsteady case a value of 0.9 was sufficient while, for the steady case, reducing those values to 0.01 and lower was crucial to have a non-diverging solution. Both pressure and temperature were limited as a function of the physics of the flow: for the pressure, the upper and lower bounds were defined using two multiplicative coefficients (100 and 0.05 respectively) and two limit temperatures were defined (10 K for the lower bound and 10000

K for the upper bound). Those values were chosen basing on what observed with similar cases studied before. The unsteady setup has an *endTime* of 0.2 with a *deltaT* of 10^{-6}; the steady setup has a limit of iteration of 30000 in function of the small Relaxation factors.
The turbulence model used is $SSTk - \omega$ with perfectGas assumption and constant transport model for the thermophysical properties. The fluid considered is air. Operating and boundary conditions have been defined according to a probable mission condition. Those have been summarized in table 1. Results have been validated both using a reference case obtained with a validate solver [4] and experimental results. Convergence criteria are based on the values of the average Dome Temperature, Drag coefficient, maximum and average values of $y+$. In table 3 are reported characteristic vales of y plus.

y^+ values		
Minimum	**Average**	**Maximum**
0.023	0.73	1.36

Table 3. Characteristic y^+ values

3. Results:

The characteristics of the base case can be seen in figure 3 where contours of the velocity magnitude are plotted. There, the presence of the disk creates a recirculating zone round the dome-shaped forebody. This results in a reduction of the local velocity magnitude that creates a reduction of the thermal load due to the viscous dissipation of the kinetic energy in the boundary layer; the stagnation point (which sees the higher thermal and pressure loads due to the combined effects of viscosity and shock) is detached from the sensible part of the vehicle thanks to the presence of the spike. The particular shape of the same induces an oblique shock that results in a reduction of the compressibility effects due to the lower strength of the wave with respect to a normal one. The combination of the spike and the recirculating zone creates a "virtual" streamlined body: because of the interaction between the freestream flow behind the disk and the recirculating zone on the dome, a second order compressibility structure is created (visible in the Schlieren image in figure 4). A computational Schlieren image has been obtained by computing the gradient of the density, which serves as a metric to look at shock waves which are a discontinuity in the density field. The effect of the aerospike is visible by comparing the cases with and without the device. Such a comparison is shown in figure 5 where contours of the static temperature is presented. From a qualitative point of view, the comparison in figure 5 shows a reduction of the compressibility effects due to the presence of a more oblique shock wave in the case with the spike and a reduction in the temperature around the dome with respect to the case without the device. This is further confirmed by the average values measured on the domeshaped surface, which are reported in table 4.

	With aerospike	Without aerospike
Temperature [K]	438	454
Drag Coeffient ratio $\frac{c_D}{c_{D,ref}}$	0.61	1

Table 4. Effects of the aerospike on the temperature and drag coefficient on the dome, with and without the aerospike.

The results show a small but meaningful reduction in the average temperature on the dome that helps thermal protection systems which are crucial to prevent heat damages to sensible parts as electronics, avionics or crew docks. An important reduction of the mean drag coefficient is achieved due to the reduction of compressibility effects which are also related to the pressure condition onto the body.

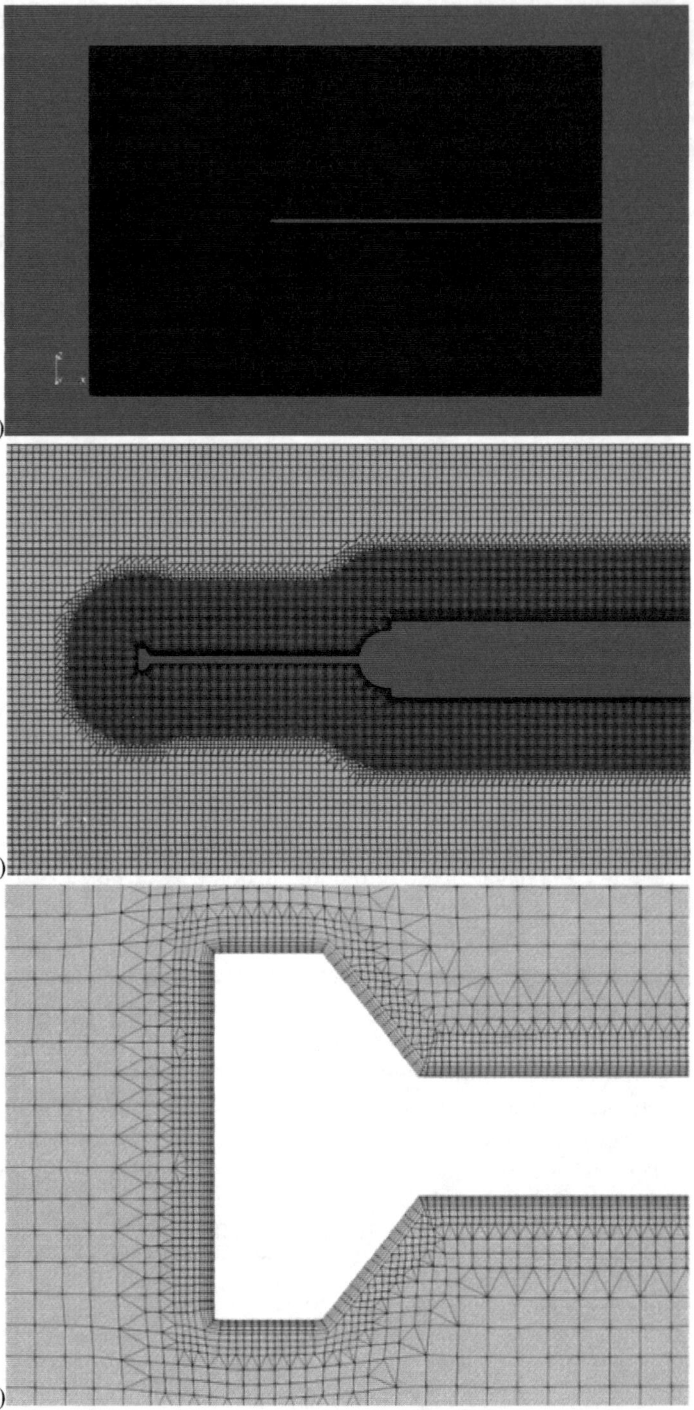

a)
b)
c)

Figure 2. aerospike mesh. (a) The fluid domain around the body, (b) The refinement region, (c) A detail of the mesh nearby the surface.

3.1. Effect due to variations of the aerospike length and disk radius.

In order to obtain an optimal aerospike design it is crucial to explore the dependency of the aerospike performances as a function of a set of variable of influence.

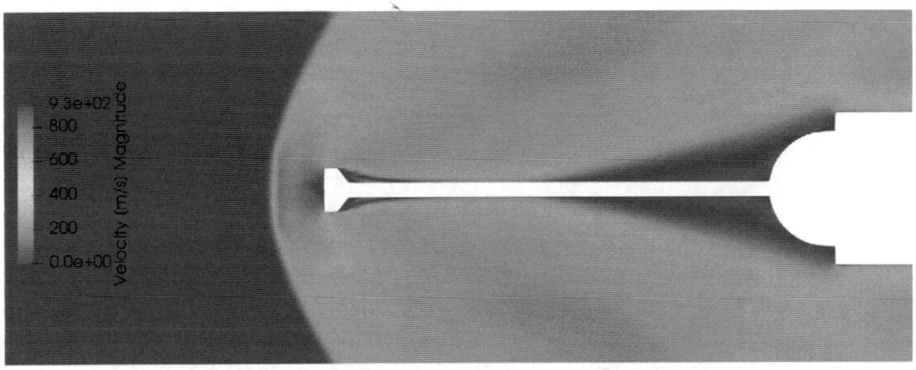

Figure 3. Velocity contour plot of the base case

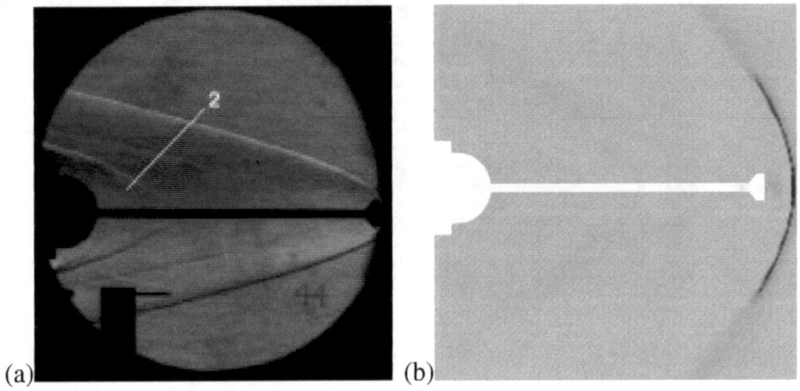

Figure 4. Comparison with experimental (a) and computational (b) Schlieren

In this particular case, those are:

- aerospike length;
- Aerodisk radius;

We first explore the dependency on the results by varying the aerodisk radius. Contour plots of the static pressure of aerodisks with three different radii are presented in figure 6.
There, it is possible to see that the increase of the disk radius promotes an increase in the flux separation afflicting the afterward recirculation zone on which relies the temperature reduction on the dome. This causes a increase in the strength of the shock wave because the flow sees, at the limit, a vertical wall moving against the flux.
The limit case with the radius of aerodisk tending to zero is presented in figure 7. In general, there is a reduction in the pressure load on the dome because the stagnation point is no more on that surface and the flux condition is driven by a flux through an oblique shock instead of a normal one. The last analysis involves changing the aerospike length. As the length decreases also the performances decreases. This is due to the fact that, fixing the radius, the detachment effect induced by the aerodisk is defined but the reduction of the spike length causes an increase of the angle that the flow sees due to the recirculating zone that appears steeper with respect to the axis of the spike. This induces a more severe condition for the compression effect due to the flow passing on the recirculation zone. In table 5 it is possible to notice this referring to the average temperature on the domeshaped forebody and the drag coefficient ratio respect to the reference value of the case without the aerospike.

(a)
(b)
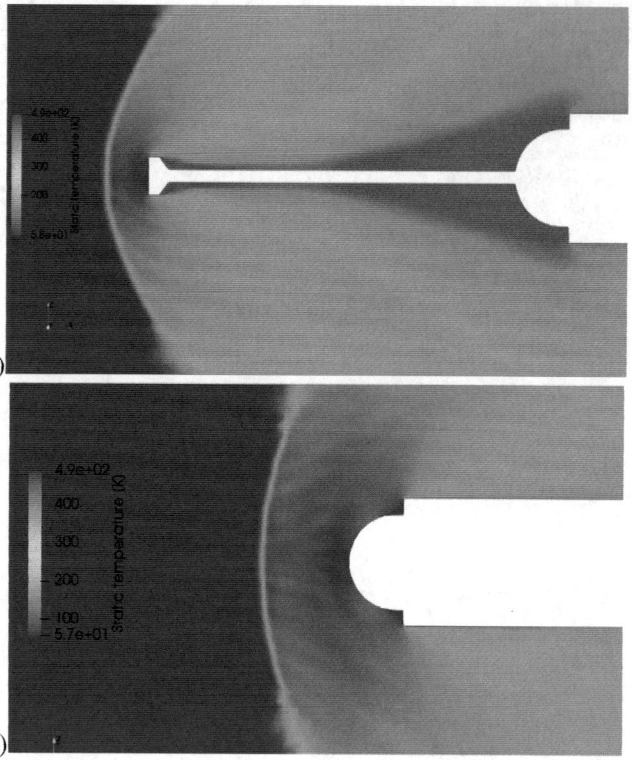

Figure 5. Static temperature contour plot: comparison with (a) and without (b) the aerospike

	Dome temperature [K]	Drag Coefficient ratio $\frac{c_D}{c_{D,ref}}$
$\frac{l}{2}$	429	0.62
l	438	0.61
$2l$	428	0.56

Table 5. Effects of the aerospike: l = reference aerospike length of the case base

4. Conclusions and future improvement:

During this work, the effectiveness of the static device called aerospike has been shown, and the effect by changing the disk radius as well as the aerospike length has been investigated. Further studies are needed to investigate different aerodisk shapes in order to find the optimal aerospike configuration.
The first step will be considering a new set of heads (cone, sphere, ...) and then try to use the adjoint method to define a new surface for the aerodisk.

From the solution method point of view there are several things that are now under investigation:

- The definition of functions to compute thermophysical properties;
- Archive a stable, converging solution using a steady solver that will help in the optimization step even thought the case appears to be unsteady;
- investigate other density based solvers;
- Define an optimization procedure with Dakota;
- Improve the mesh quality to better follow the flux evolution.

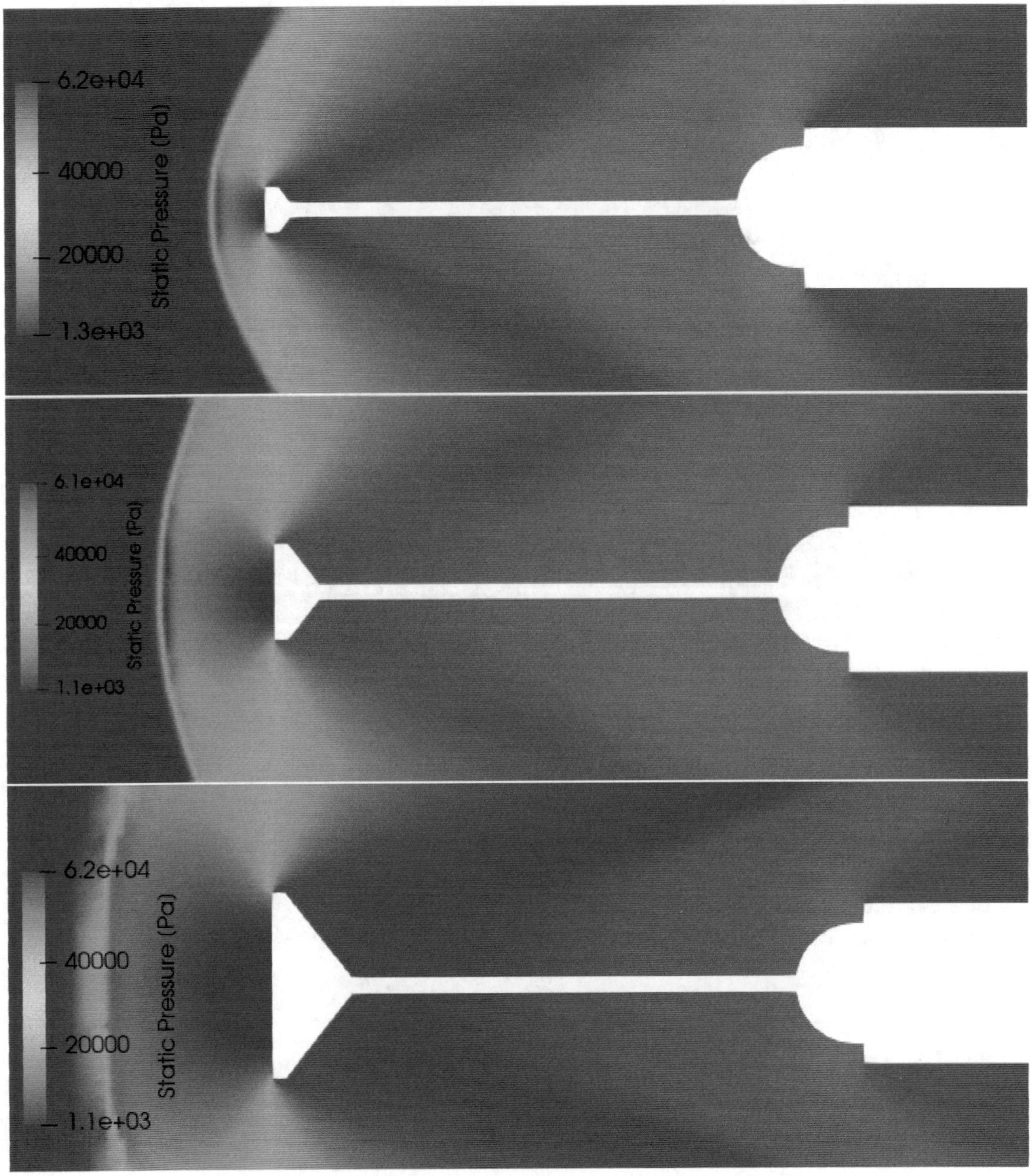

Figure 6. Static pressure contour plot: effect of the radius

(a)
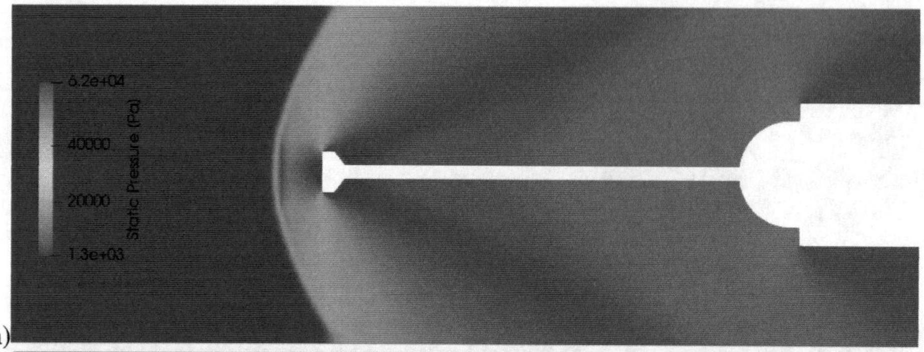

(b)
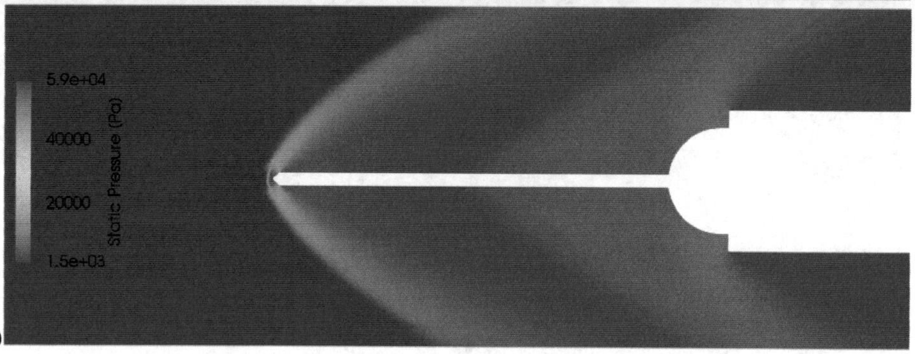

Figure 7. Static pressure contour plot: with (a) and without (b) the aerodisk

Availability of data and materials Cases (including meshes), comparison data, and calculation results presented in this paper are available from the corresponding author upon reasonable request.
Geometries and case setup folder are available at:
https://github.com/R0Carbone/AeroSpike

Acknowledgement
The authors acknowledge the organizing committee of the 18th OpenFOAM Workshop for the support provided.

References

[1] M. B. Gerdroodbary and S. Hosseinalipour, "Numerical simulation of hypersonic flow over highly blunted cones with spike," *Acta Astronautica*, vol. 67, no. 1, pp. 180–193, 2010.

[2] P. Gnemmi, J. Srulijes, K. Roussel, and K. Runne, "Flowfield around spike-tipped bodies for high attack angles at mach 4.5," *Journal of Spacecraft and Rockets*, vol. 40, no. 5, pp. 622–631, 2003.

[3] L. D. Huebner, "Experimental Results on the Feasibility of an Aerospike for Hypersonic Missiles," *33rd Aerospace Sciences Meeting and Exhibit*, vol. AIAA 95-0737, 1995.

[4] Carbone, R and Innocenti, G., "Aerospike," *Università di Genova, Turbulence and Optimization courses. Final Project*, 2022.

IOP Conf. Series: Materials Science and Engineering **1312** (2024) 012003
doi:10.1088/1757-899X/1312/1/012003

Evaluation of a carbon dioxide fish barrier with OpenFOAM

M Politano[1], A Cupp[2], D Smith[1], A Schemmel[1], P R Jackson[3] and J Zuercher[4]

[1] Environmental Laboratory, US Army Engineer Research and Development Center, Vicksburg, MS 39180-6199 USA
[2] US Geological Survey, Upper Midwest Environmental Sciences, La Crosse, WI 54603 USA
[3] US Geological Survey, Central Midwest Water Science Center, Urbana, IL 61801 USA
[4] US Army Corps of Engineers, Chicago District, Chicago, IL 60604 USA

Marcela.s.politano@erdc.dren.mil

Abstract. The Chicago Area Waterway System (CAWS) is a potential route for the migration of aquatic invasive species from the Mississippi River basin into the Great Lakes. Electric deterrence barriers were constructed in the Chicago Sanitary Ship Canal (CSSC), within CAWS, to prevent invasive fish from reaching the Great Lakes. Despite the high efficiency of these barriers, occasional maintenance events are a weakness in the system that fish can exploit to access the Great Lakes. This study aimed to assess the feasibility of a carbon dioxide (CO_2) infusion system to deter fish during the maintenance of the electric barriers. A simplified two-fluid model was implemented in the OpenFOAM solver to represent the underwater CO_2 bubble plume and predict the concentration of dissolved CO_2 in the canal. Simulations under three canal flowrates and two sparger systems were conducted assuming a constant gas flowrate. Numerical results indicate that, for all simulated conditions, the CO_2 concentration is not fully mixed creating passageways that invasive fish could potentially use to migrate upstream. Injecting 4-mm bubbles induces two large-scale recirculations that are expected to synergistically improve fish avoidance. On the other hand, injection of 20 μm bubbles results on almost immediate dissolution with minimal impact on the flow pattern. To improve effectiveness, a pulse system was proposed to create a CO_2 gradient, and thus promote a more responsive behaviour from fish. According to the simulations, this system not only increases efficiency but also extends the operation of the CO_2 barrier. Moreover, the pulse concept mitigates potential impact of elevated CO_2 downstream of the CSSC.

1. Introduction

Invasive species pose a major threat to the Great Lakes, one of the United States' most vital freshwater resources. Of particular concern is the presence of Invasive carps, which have dramatically increased in the Mississippi basin, and disrupted ecosystems by altering habitats and destabilizing the food chain [1][2][3].

The Chicago Area Waterway System (CAWS) spans over 100 miles of river and channels, connecting the Mississippi River with Lake Michigan. The Chicago Sanitary and Ship Canal (CSSC), within the CAWS, is an artificial shipping channel that provides a connection for fish passage [2][4]. To prevent the invasive fish from reaching the Great Lakes through the CSSC, the U.S. Army Corps of Engineers (USACE) constructed and operates an electric dispersal barrier system. These barriers have proved to be efficient fish deterrence [5][6]. However, the effectiveness of the electrical barriers is compromised during occasional maintenance events when they must be turned off. This temporary shutdown creates a vulnerability in the deterrence system, creating an opportunity for invasive fish to exploit and gain access to the Great Lakes.

Recent research indicates avoidance behaviour of invasive species to elevated carbon dioxide (CO_2) concentrations ([7][8][9][10][11][12]). Ongoing research is being conducted to fully understand deterrence of Invasive carps and other invasive species using CO_2. The specific CO_2 concentration that elicits a reaction from fish can vary depending on the species, size and life stage, duration of exposure and environmental conditions. Three hours of exposure in a laboratory to CO_2 concentrations of 30 ppm had minimal physiological or behavioural impacts on adult carps, but the same time exposure to 70 ppm resulted in a drop in ventilation rates and irregular behaviours [7]. A CO_2 concentration of approximately 60 ppm is generally effective in eliciting an avoidance response from most carps [12]. The U.S. Environmental Protection Agency (USEPA) pesticide label for Carbon Dioxide-Carp allows up to 100-150 mg/L to deter invasive carp species.

In this study, a large-scale CO_2 infusion system located within the existing electric barrier area was proposed for deployment during the maintenance of the barrier. The system comprises diffusers that generate a plume of bubbles, which dissolve into the water. A consistently high CO_2 concentration may potentially overwhelm fish, causing them to hunker down rather than move out of the electrical barrier reach. Based on fish response to other stimuli, a CO_2 concentration gradient is expected to provide fish a refuge from the highest CO_2 levels directing them downstream and thereby increasing the system's efficiency as a deterrence. In this study, two infusion concepts were evaluated. The first is a constant gas flowrate and the second a CO_2 pulse system as illustrated in Figure 1. A maximum target concentration of 150 ppm was selected to ensure a satisfactory fish avoidance response.

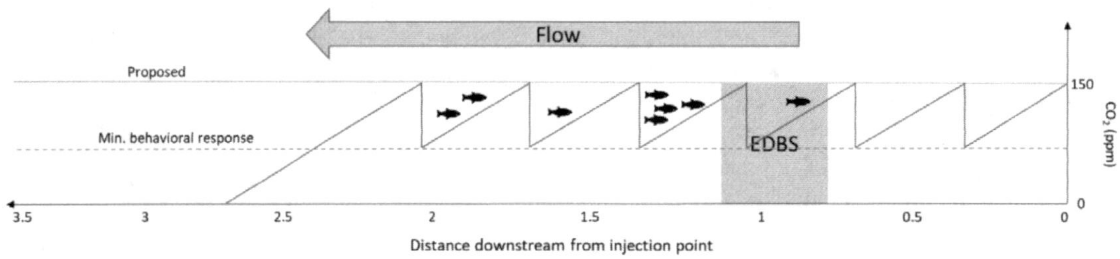

Figure 1. CO_2 pulse concept.

Advancements in computational power, expanded memory and increased storage have enabled the use of computational fluid dynamics (CFD) to model complex two-phase flows. OpenFOAM is an open-source CFD software package widely recognized for its flexibility, robustness, and versatility [13]. In this study, a simplified two-phase flow model, neglecting the bubble inertia, was implemented in the OpenFOAM solver buoyantBoussinesqPimpleFoam. The main advantage of this model is its stability and speed, achieved through the solution of a system of algebraic equations rather than differential equations for the bubble velocity.

The aim of this paper is to assess the feasibility of the CO_2 infusion system in deterring the migration of Invasive carps in the CSSC. Numerical simulations were conducted for three canal flowrates and two bubble sizes at the injection point that represented expected operation scenarios at this location. The performance of the system was assessed by delineating deterrence regions in the barrier near-field area and examining changes in the hydrodynamics resulting from the injection of the bubble plume. In addition, the study examined the downstream transport and dispersion of elevated CO_2 concentrations with the goal of evaluating potential exposure risks to non-target organisms located outside of the treatment reach.

2. Study area

The study area comprises a stretch of 18.5 km along the CSSC, extending from the USGS stream gage 05536890 near Lemont, IL to Lockport Lock and Dam. The electric barriers are located near Romeoville, IL.

The bathymetry used by [14] was incorporated into the CFD model. The authors combined bathymetric data surveyed by the USGS in 2009 and data collected by USACE in 2008. Figure 2 shows the model extension and bathymetry used in this study, referenced to the North American Vertical Datum of 1988 (NAVD88).

Figure 2. Model extension and bathymetry.

The proposed infusion system consists of four pipes placed into the existing electric barrier area at the bottom of the canal. Each of the pipes is equipped with 400 μm fine bubble diffusers. The CO_2 gas introduced forms a plume of small bubbles that dissolve in water.

Approximately 1.8 km downstream of the infusion system lies the Commonwealth Edison coal-fired power plant. The plant, currently in the process of closing operation, operates as a once-through cooling system, drawing water from the CSSC and returning warm water downstream.

3. Modelling bubble plumes

A bubbly flow, characterized by a discrete gas phase in a continuous fluid, is formed in a bubble plume. Bubbles injected into a river or canal induce a buoyancy-induced flow, driven by the density difference between the bubbles and the liquid ([15][16][17][18][19]). The model used in this study is valid for the plume zone, where most of the dissolution occurs. The model assumes that bubbles reached a pseudo-steady state immediately after release, excluding the jet zone. Measurements of gas volume fraction, bubble frequency and chord length indicate that bubble breakup predominantly occurs near the injection location [19]. Close to the free surface, the bubble velocity decreases as liquid moves tangent to the free surface, enhancing coalescence and leading to larger bubble sizes. The model neglects this effect as the infusion system is designed to dissolve most bubbles before reaching the free surface.

This study adopts a rigid-lid approach that prescribes the free surface. This method is computationally more efficient than computing the free surface, but it does not account for the free surface deformation by the bubble plume. Bubbles at the injection point are represented with a single characteristic bubble size. After injection, bubbles size changes by dissolution and pressure neglecting breakup and coalescence.

Three types of models have been used to model bubbly flows: 1) Euler-Lagrangian models, 2) two-fluid models, and 3) quasi single-phase models. In Euler-Lagrangian models the liquid phase is treated as a continuum, which can be solved with the Navier-Stokes equations similar to a single-phase flow. The motion of the discrete bubbles is tracked using a Lagrangian approach considering interfacial forces such as drag, lift and buoyant forces. Forces exerted by the bubbles on the liquid phase can be included in the momentum equations as source terms. Most of the Lagrangian approaches neglect bubble-bubble interaction and therefore the method is only valid for low gas volume fractions. A comprehensive review of gas–liquid flow simulations in bubble columns using the Eulerian–Lagrangian approach is found at [20]. Former studies have shown that Lagrangian models can accurately predict the flow induced by bubbles [21][22][23][24][25]. However, the computational resources required to track the high number of bubbles required in a large-scale bubble plume limits the application of these models. In two-fluid models, the disperse phase is treated as a second continuous phase that interacts with the continuous liquid phase [26]. The model is obtained by ensemble averaging of the mass and momentum conservation equations for each phase. Interaction terms representing exchange of mass and momentum between the phases need to be modelled. Most simulations of bubble plumes using a two-fluid model focus on laboratory experiments conducted at a relatively small scale. Quasi single-phase models consider a liquid phase with a reduced density in the zones where bubbles are present. While two-fluid models can be applied to large-scale flows, most studies found in the literature on bubble plumes in rivers or channels used a quasi single-phase flow approach due to the large computational resources required by two-fluid models. Bubble plume simulations were mostly applied to simple geometries where the bubble zone is assumed to have a given geometrical shape [16][27][28]. A mathematical model for large-scale bubble plumes based on the two-fluid model using the concept of mixture phase and the Boussinesq approach was proposed by [29]. The authors assumed a constant bubble size and used an expression for the bubble relative velocity as a function of the radius. This model was developed as an alternative to the more comprehensive two-fluid models. The main advantage of this model is that an equation for the gas phase is included and therefore the shape of the bubble zone can be predicted. Future model improvements such as inclusion of a bubble size distribution, breakup and coalescence, bubble dissolution, bubble compression, bubble induced turbulence, etc., can easily be incorporated into the model using this formulation.

The model used in this study for the CO_2 bubble plume in the CSSC accounts for the relevant physics including buoyancy, drag, turbulent dispersion, and CO_2 dissolution. The buoyancy-induced flow by the bubbles was modelled following the mixture concept and Boussinesq approach proposed by [29]. An equation for the bubble number density was included to estimate bubble size changes by dissolution and compression. The gas-liquid relative velocity was computed using the momentum equation for the gas phase.

4. Mathematical model

4.1. Mixture model

The algebraic slip mixture model is a simplified two-fluid model that can be obtained adding the mass and momentum equations for each phase of the two-fluid model. Conservation equations for mass and momentum of the mixture air-water, assuming that density variations are small enough that have no effects on the flow field except on the buoyancy force (Boussinesq approach) are [29]:

$$\nabla \cdot \overrightarrow{U_m} = 0 \tag{1}$$

$$\frac{\partial \overrightarrow{U_m}}{\partial t} + \nabla \cdot \left(\overrightarrow{U_m}\,\overrightarrow{U_m}\right) = -\nabla\left(\frac{P}{\rho_m}\right) + \nabla \cdot \left(\frac{\tau_m^{Re} + \tau_m}{\rho_m}\right) + \overrightarrow{f_g} = 0 \tag{2}$$

where $\vec{U}$ is velocity, ρ is density, τ_m^{Re} and τ_m are turbulent and molecular stress tensor, and P is pressure. Subscripts l, g and m denote the liquid, gas and mixture phase, respectively. $\overrightarrow{f_g}$ is the buoyancy force, which can be modelled as:

$$f_g = -\frac{(\rho_l - \rho_g)}{\rho_m}\alpha\,\vec{g} \tag{3}$$

where α is the gas volume fraction and $\vec{g}$ the gravity.

4.2. *Gas phase*

The continuity equation for the gas phase is:

$$\frac{\partial \rho_g \alpha}{\partial t} + \nabla \cdot \left(\overrightarrow{U_g}\,\rho_g\,\alpha\right) = \nabla \cdot \left(\left(\nu + \nu_b + \frac{\nu_t}{Sc}\right)\nabla(\rho_g \alpha)\right) - S \tag{4}$$

where S is the bubble dissolution, Sc the Schmidt number, ν and ν_t are the molecular and turbulent kinematic viscosity, respectively. ν_b represents the turbulence generated by the bubbles. The gas density ρ_g is computed assuming ideal gas law and hydrostatic pressure.

The momentum equation for the gas phase considering that inertia, gravity force and viscous shear stresses are significantly smaller than liquid-gas interfacial forces due to the small density and viscosity of the gas phase is ([30][31][32]):

$$0 = -\alpha\,\nabla P - \frac{3}{4}\alpha\frac{C^D}{D_g}\overrightarrow{U_r}\left|\overrightarrow{U_r}\right| \tag{5}$$

where $\overrightarrow{U_r} = \overrightarrow{U_g} - \overrightarrow{U_l}$ is the gas-liquid relative velocity, D_g the bubble diameter and C^D the drag coefficient, which can be modelled following [30].

4.3. *Bubble size*

Two approaches are possible for modelling bubble size. The more comprehensive one uses a polydisperse model which assumes that bubble size distribution can be represented with groups of constant mass [33]. A simplified approach to take into account size changes by compression and mass transfer is possible assuming one variable bubble size (instead of a bubble size distribution) at each time and position, as used by [34].

In the free bubble zone, it can be assumed that a steady-state, fully developed flow condition is reached, and the equilibrium of bubble breakup and coalescence results in a given bubble size distribution. The transport equation for the bubble number density N results:

$$\frac{\partial N}{\partial t} + \nabla \cdot \left(\overrightarrow{U_g}N\right) = \nabla \cdot \left(\left(\nu + \nu_b + \frac{\nu_t}{Sc}\right)\nabla N\right) \tag{6}$$

The bubble diameter is calculated from $D_g = \left(\frac{6\,\alpha}{\pi N}\right)^{\frac{1}{3}}$.

4.4. *Dissolved carbon dioxide concentration*

The transport equation for the carbon dioxide concentration is:

$$\frac{\partial C}{\partial t} + \nabla \cdot \left(\overrightarrow{U_l}\,C\right) = \nabla \cdot \left(\left(\nu + \nu_b + \frac{\nu_t}{Sc}\right)\nabla C\right) + S \tag{7}$$

The rate of mass transfer between bubble and liquid phases, considering that CO_2 obeys Henry's law, is modelled as:

$$S = \pi N\, D_g^2\, k_l \left(\frac{P}{H} - C\right) \tag{8}$$

where H is the Henry constant and k_l the mass transfer coefficient, which can be calculated with the expression proposed by [35].

4.5. Turbulence closure

For a bubble curtain, an isotropic turbulence model as the $k - \varepsilon$ model can be used for turbulence closure. The mixture model approach assumes that gas and liquid phases share the same turbulence field. The turbulence in the mixture phase is computed using the transport equations for a single phase but with properties and velocity corresponding to the mixture. A simplified approach to consider the effect of the bubbles on the turbulence field is to define a total effective viscosity as a linear superposition of the liquid viscosity and the bubble-induced turbulent viscosity, which can be modelled as [36]:

$$\nu_b = 0.6 D_g\, \alpha\, U_r \tag{9}$$

5. Numerical model

The model used in this study is based on the open source CFD code OpenFOAM v2112. The code uses a finite volume method, adopting a collocated methodology on an unstructured polyhedral grid with arbitrary grid elements. Specifically designed for heat transfer problems, the transient solver buoyantBoussinesqPimpleFoam solves the discrete Reynolds-Averaged Navier-Stokes (RANS) for turbulent flow of incompressible fluids using the Boussinesq approximation.

In this study, a new solver, buoyantBoussinesqMixPimpleFoam, was developed. In addition to temperature, the solver computes the gas field, incorporates the buoyancy force induced by the bubbles in the liquid phase, and solves dissolved CO_2 in the liquid phase (Eqs. (4) to (13)). It is assumed that bubbles enter the domain at their terminal velocity.

The buoyantBoussinesqPimpleFoam solver defines the buoyancy force due to temperature, f_T, as:

$$f_T = \left[1 - \beta\left(T - T_{ref}\right)\right] \vec{g} \tag{10}$$

where T is temperature and β the thermal expansion coefficient. T_{ref} is the reference temperature. OpenFOAM solves the momentum equation via the piezometric pressure combining pressure and gravity:

$$P_{rgh} = P - \rho_k\, \vec{g}.\nabla z \tag{11}$$

with $\rho_k = \left[1 - \beta\left(T - T_{ref}\right)\right]$. The buoyantBoussinesqMixPimpleFoam solver incorporates the buoyancy by the bubbles given by Eq. (3) resulting in:

$$\rho_k = \left[1 - \beta\left(T - T_{ref}\right)\right](1 - \alpha) + \frac{\rho_g}{\rho_l}\frac{T}{T_{ref}}\, \alpha \tag{12}$$

In OpenFOAM equations are solved separately for each variable at each time step with an iterative procedure (segregated method). The PIMPLE algorithm solves iteratively the momentum equations enforcing the incompressibility constrain. It is a combination of PISO (Pressure Implicit with Splitting of Operator) [37] and SIMPLE (Semi-Implicit Method for Pressure-Linked Equations) [38]. Figure 3 shows a flowchart of the solver.

The CSSC model uses non-orthogonality correction for stability and accuracy.

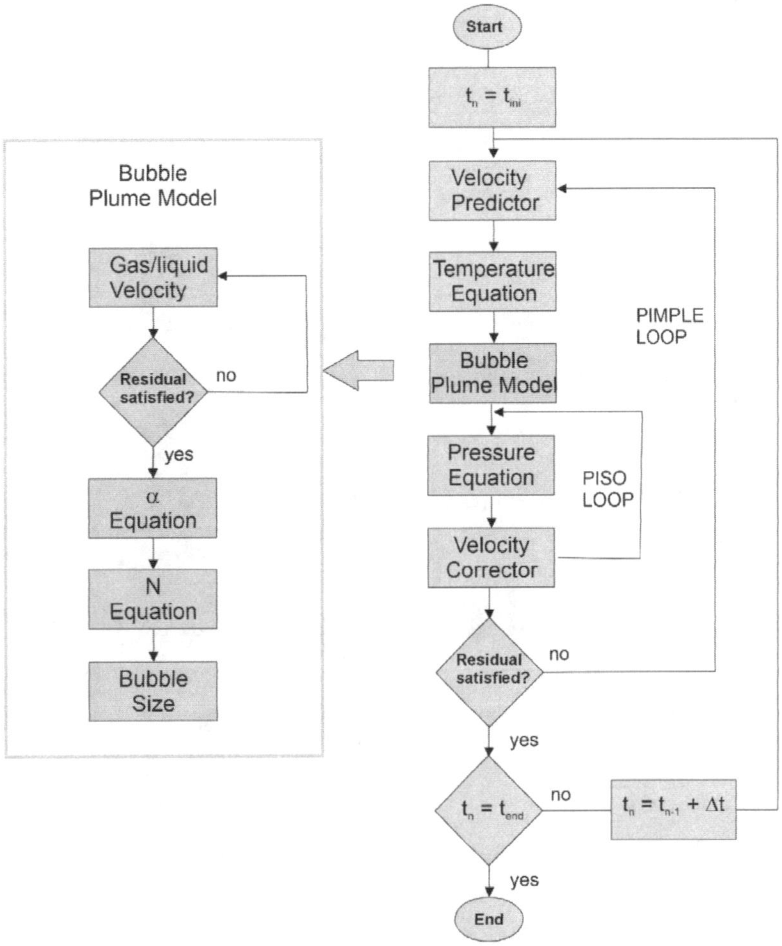

Figure 3. Flowchart of buoyantBoussinesqMixPimpleFoam.

6. Validation of the two-phase flow model

The model was used to simulate a water-filled cylindrical tank agitated by a gas bubble stream and velocity data was compared against measurements inside the tank [16]. The modelled tank radius, R, was 0.3 m and the water height, H, 0.6 m. Bubbles were injected into the quiescent liquid through an orifice of 0.0127 m at a flowrate of 205 m^3/s.

The numerical grid was generated with the commercial grid generator Pointwise. Grid size was approximately 10^5 nodes. Figure 4 (left) shows the grid and boundary conditions. Only one fourth of the tank was simulated to reduce grid size and computational time and symmetry boundary conditions were used on the sides. The images in the top and bottom show results at the symmetry plane and top of the model, respectively. Boundary condition types for the sparger, outflow and walls are shown in Table 1.

Figure 4 (right) shows the gas volume fraction distribution in the tank. For the low gas volume fraction injected, bubbles concentrate in the core of the tank and rise almost uniformly in a nearly straight line. Near the sparger, bubble velocity increases due to the upward liquid flow in the centre of the tank. This local increase of the liquid velocity causes a reduction of the gas volume fraction. Conversely, the tangential liquid flow at the free surface results in a local increase of the gas volume fraction.

Reasonable agreement was found between model predictions and experimental data for three axial positions (Figure 5). Using a bubble size distribution at the inlet is expected to improve model accuracy. This approach is proposed for future model improvement once measurements of the bubble size distribution near the sparger become available. The model is considerably less accurate near the walls.

IOP Conf. Series: Materials Science and Engineering **1312** (2024) 012003 doi:10.1088/1757-899X/1312/1/012003

The incorporation of the wall lubrication force together with the lift force is essential for reproducing the gas volume fraction near the wall [33]. However, since wall interactions with the bubble plume are not expected to play a significant role on the application presented in this paper, wall interfacial forces were not included in the model.

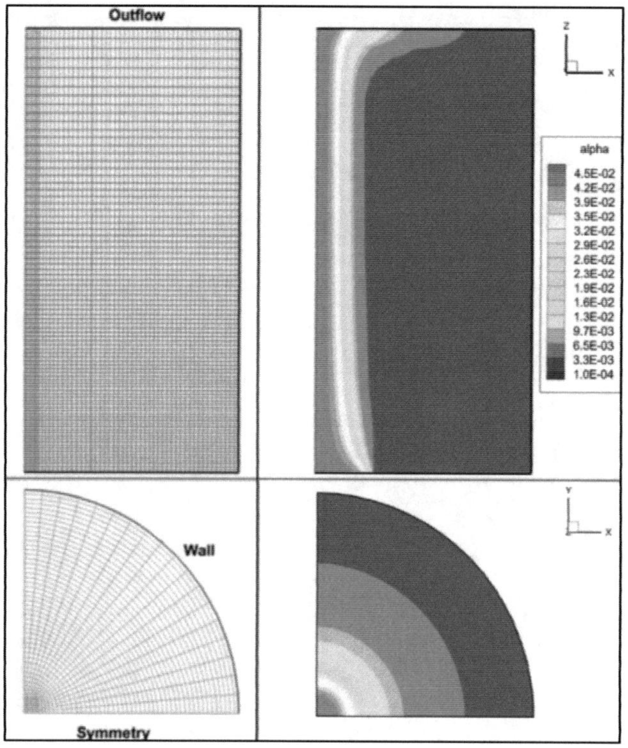

Figure 4. Grid and boundary conditions (left) and gas volume fraction (right).

Table 1. OpenFOAM boundary conditions used for the validation and the CSSC model. calc: calculated, eWF: epsilonWallFunction, fRIV: flowRateInletVelocity, fV: fixedValue, iO: inletOutlet, kRWF: kqRWallFunction, nkWF: nutkWallFunction, nS: no slip, S: slip, tIKEI: turbulentIntensityKineticEnergyInlet, zG: zeroGradient

Tank	--	Walls	--	Diffuser			
CSSC	Inlet	Riverbed/ Walls	Outlet	Diffuser	Powerplant Inflow	Powerplant Outflow	Free surface
U	fRIV	nS	iO	nS	fRIV	iO	S
p	zG	zG	fV	zG	zG	fV	S
k	tIKEI	kRWF	iO	tIKEI	tIKEI	iO	zG
ε	fV	eWF	zG	eWF	fV	zG	zG
CO_2	fV	nS	zG	nS	fV	zG	
α	fV	nS	zG	nS	fV	zG	zG
N	fV	nS	zG	nS	fV	zG	zG
nut	calc	nkWF	zG	NWF	calc	zG	zG

IOP Conf. Series: Materials Science and Engineering **1312** (2024) 012003 doi:10.1088/1757-899X/1312/1/012003

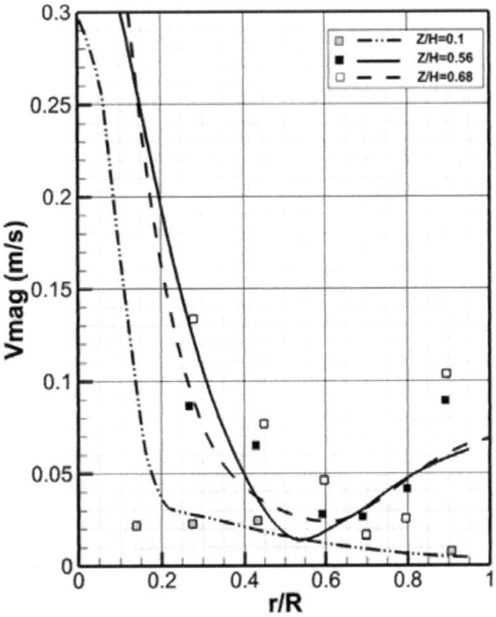

Figure 5. Comparison between predicted (lines) and measured (symbols) velocity magnitude.

7. CSSC model

7.1. Numerical grid and boundary conditions

The numerical grid of the CSSC model was generated using the OpenFOAM mesh generator snappyHexMesh, which creates 3D meshes containing hexahedra and split-hexahedra automatically from triangulated surface geometries in Stereolithography (STL) format. The mesh conforms to the STL surface by iteratively refining a starting Cartesian grid and morphing the resulting split-hex mesh to the surface. The stl file with the bathymetry of the CSSC was created with the USACE Surface-water Modeling System (SMS) software.

Boundary conditions and grid details with typical grid dimensions are shown in Figure 6. The water through the power plant intake was modelled as an outflow and the flowrate from the plant was considered an inflow. The top of the grid was modelled assuming a slip boundary condition. CO_2 gas was injected using four feed system, each with an area of approximately 37.1 m^2. Table 1 describe the type of OpenFOAM boundary condition used for each variable of the model.

A background mesh was generated with the OpenFOAM utility blockMesh. The aspect ratio of the background mesh was 2.5 to avoid an excessive grid size and hence computation time. The region 1 km upstream of the sparger was refined to properly predict the velocity profile and turbulence characteristics approaching the diffuser. The grid was also refined approximately 2.6 km downstream of the diffuser to capture the effect of the bubble plume on the flow field and CO_2 distribution in the area of interest. Table 2 shows dimensions of the internal cells and boundaries. The grid size of the model was 3.3 10^6 cells.

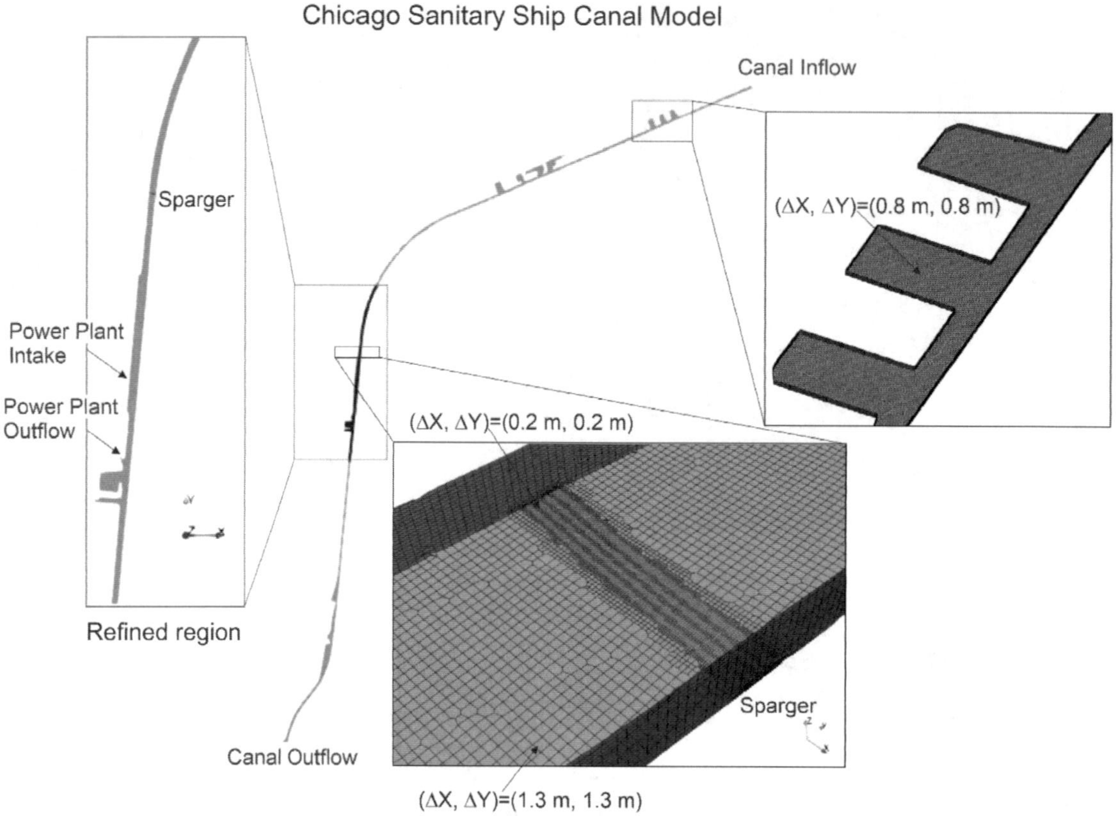

Figure 6. Modelled domain with refined region.

Table 2. Typical mesh element size (m) in the longitudinal, lateral and vertical directions.

	Internal Cells	**Inflows/Outflows/Side/Sparger**	**Riverbed**
Coarse region	(5.3 5.3 2.1)	(1.3 1.3 0.5)	(2.6 2.6 1.1)
Refined region	(3.5 3.5 2.1)	(0.2 0.2 0.1)	(0.8 0.8 0.5)

7.2. Simulation conditions

7.2.1. Constant CO_2 gas feeding. Simulations under three canal flowrates corresponding to 15%, 50% and 90% flow duration measured at the USGS gauge station near Lemont were performed to evaluate the feasibility of a constant gas feeding system to produce a uniform CO_2 concentration distribution in the CSSC. Table 3 summarizes simulation conditions. Two bubble sizes at the injection point, 4 mm and 20 μm, were simulated to evaluate the impact of the sparger type on the dissolved CO_2 concentration distribution. The top of the grid was placed at 575.4 NAVD88 (water depth at the diffuser region about 8 m) corresponding to the average water surface elevation at Lockport Lock and Dam from Jun. 2019 to Dec. 2020.

A typical water withdrawal for the power plant cooling of 53.8 m^3/s was used in the simulations with 15% and 50% flow duration. For the 90% canal flow duration, plant withdrawal was assumed 46.7 m^3/s. Water residence time in the power plant was 5 min. It was assumed that CO_2 concentration does not change in the power plant and therefore water from the outfall enters the canal with the mass flow average CO_2 concentration at the intake location.

IOP Conf. Series: Materials Science and Engineering **1312** (2024) 012003 doi:10.1088/1757-899X/1312/1/012003

All simulations assumed negligible wind at the free surface. A constant CO_2 gas feed rate of 8.5 kg/s was use for all simulations.

Simulations were run without bubbles until a steady state condition was reached. The initial condition for CO_2 was the ambient Lake Superior CO_2 concentration of 16.4 ppm.

Table 3. Simulation conditions and dissolved CO_2 concentration

	Flowrate (m^3/s)	Flow duration (%)	Bubble size (mm)	Gas/water ratio (ppm)	Bubble dissolution (%)	CO_2 concentration (ppm)
Sim1	83.25	50	4	102.1	89	90.9
Sim2	83.25	50	0.02	102.1	100	102.1
Sim3	127.42	90	4	66.7	100	66.7
Sim4	127.42	90	0.02	66.7	100	66.7
Sim5	52.95	15	4	160.5	80	128.4
Sim6	52.95	15	0.02	160.5	100	160.5

7.2.2. Pulsing CO_2 gas feeding. A simulation was performed for the canal flowrate corresponding to 50% flow duration and bubble diameter 4 mm to evaluate the feasibility of a pulse system to produce a CO_2 concentration gradient able to deter fish from the barrier area. Run time of the CO_2 system was 17.7 min. and time off was 11.75 min. The Romeoville powerplant was considered non-operational for this simulation.

7.3. Simulation results

7.3.1. CO_2 distribution and flow pattern near the infusion system for a constant gas injection. Figure 7 shows the distribution of CO_2 gas volume fraction and flow pattern in a vertical slice passing through the infusion system for the simulations with injected 4 mm bubble size. Figure 8 shows the dissolved CO_2 concentration near the bubble plume at the same location. The top, middle and bottom frames show results for Sim1, Sim3 and Sim5, respectively (50%, 90% and 15% flow duration). Bubbles generated in the diffuser rise to the free surface due to the buoyancy force and are transported downstream of the injector by the water flow in the canal. Bubbles dissolve in water increasing CO_2 concentration as they move in the water column. The downstream distribution of dissolved CO_2 depends on the water velocity and turbulence level.

The effect of the bubble plume on the flow pattern is noticeable, particularly at low canal flowrates. As rising bubbles leave the sparger, they attract water near the canal bed toward the plume. The flow pattern is driven by the rising bubbles as well as secondary currents created by the canal curvature upstream of the diffuser.

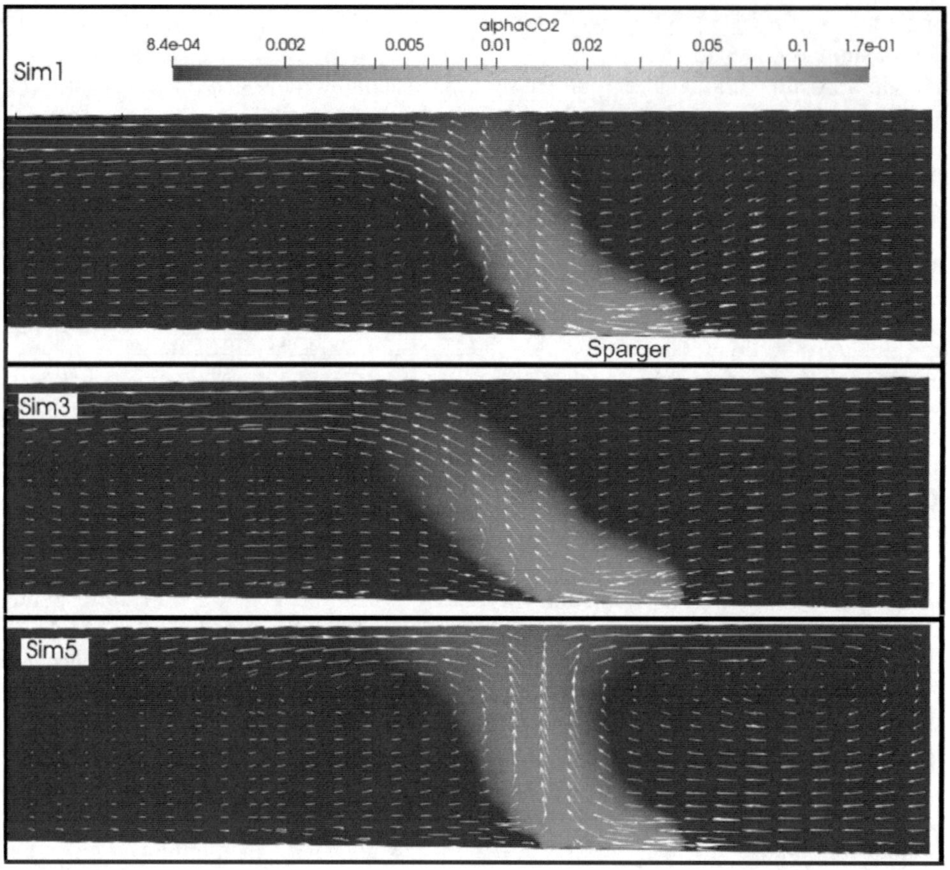

Figure 7. CO_2 gas volume fraction and velocity vectors for 4 mm bubbles injection.

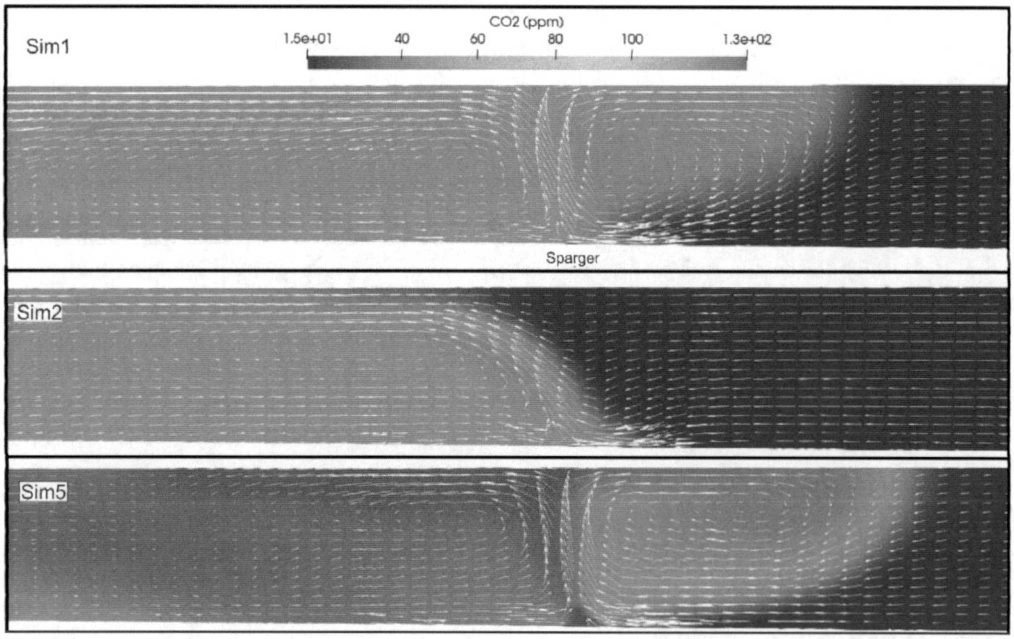

Figure 8. Dissolved CO_2 concentration and velocity vectors for 4 mm bubbles injection.

Figure 9 shows CO_2 isosurfaces near the infusion system. For all simulation conditions with injected 4 mm bubbles, CO_2 distribution is not uniform and invasive fish could find passageways to migrate upstream.

Figure 10 shows the distribution of dissolved CO_2 and flow pattern in a vertical slice passing through the infusion system for the simulations with injected 20 μm bubbles. The large interfacial area of the bubbles result on almost immediate dissolution after injection with minimal impact on the flow pattern, irrespectively of the canal flowrate. Elevated CO_2 concentrations are observed near the canal bottom with minimal vertical mixing. It is important to note that the current model neglects the turbulence generation at the infusion system as well as the expansion of bubbles released from the highly pressurized system, and bubble breakup and coalescence. All the above processes are expected to increase mixing and improve the efficiency of the system.

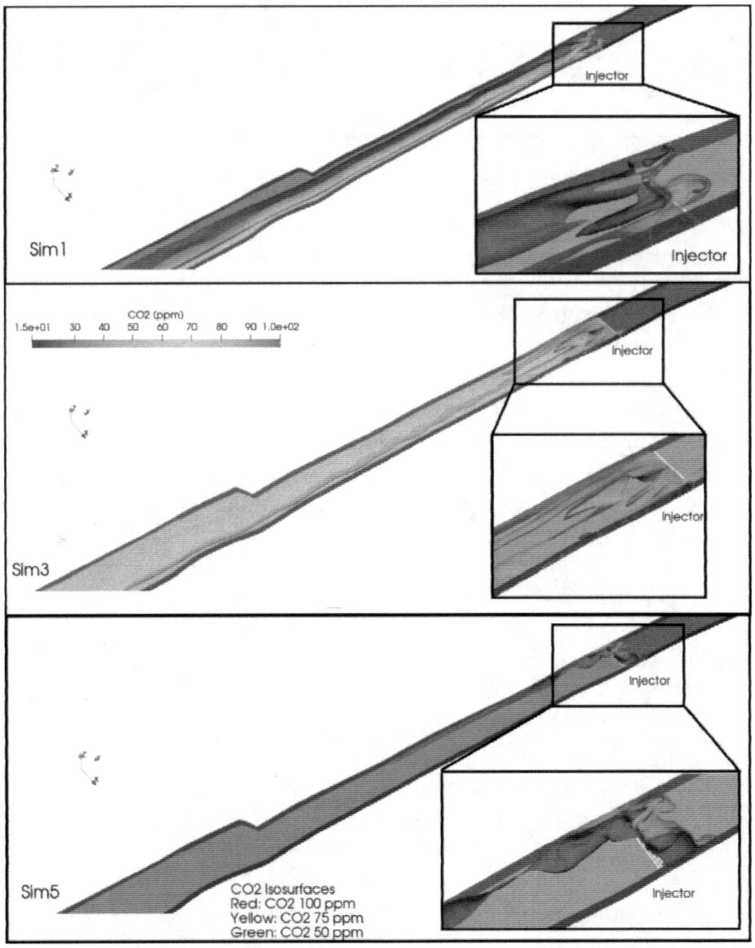

Figure 9. Dissolved CO_2 isosurfaces for 4 mm bubbles injection.

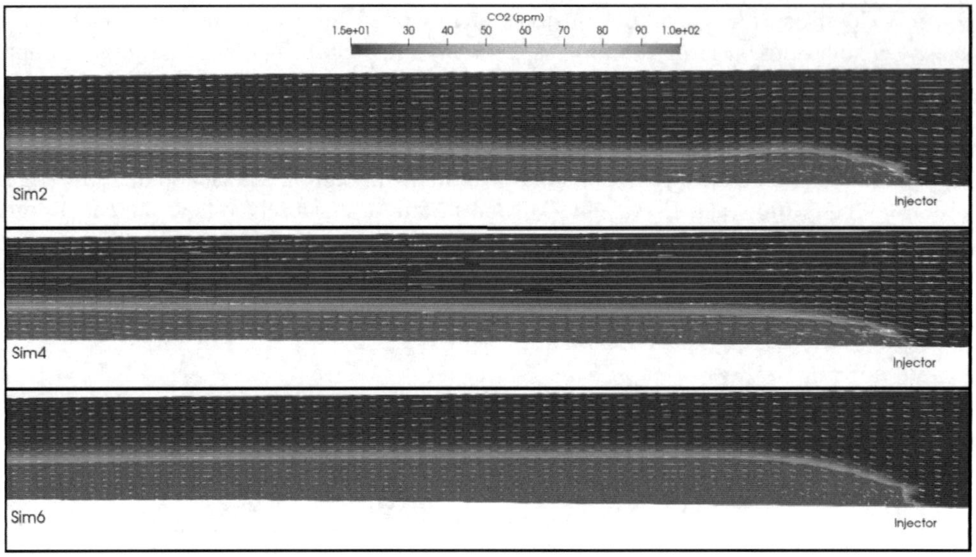

Figure 10. CO_2 concentration and velocity vectors for 20 μm bubbles.

As shown in Table 3, the average CO_2 concentration in a cross section downstream of the injector for the 15%, 50% and 90% flow durations, with an injection of 4 mm bubbles, is 66.7 ppm, 90.9 ppm and 128.4 ppm, respectively. All bubbles dissolve in water for the 15% flow duration (Sim3). On the other hand, at the 50% (Sim1) and 90% (Sim5) flow durations, bubbles leave the free surface before complete dissolution. Nevertheless, the average CO_2 concentration in a cross section downstream of the sparger decreases with the flowrate as the amount of gas injected per unit volume of water is smaller than the difference in bubble dissolution. Furthermore, increased turbulence at higher canal flowrates enhances bubble dissolution. On the other hand, full dissolution of the 20 μm bubbles results in a higher average CO_2 concentration (see Table 3). However, it is important to note that CO_2 concentrations targeting an elicit response are only near the bottom of the canal, and the system with the 20 μm bubbles is less efficient at acting as a barrier, especially if invasive fish swim near the free surface.

7.3.2. Downstream CO_2 distribution far from the infusion system for a constant gas injection. Figure 11 shows the dissolved CO_2 concentration in a horizontal slice at 1.5 m from the free surface for Sim3. Water with elevated CO_2 concentration is transported 8.2 km downstream of the barrier to Lockport Lock and Dam. The time taken to transport the CO_2 plume from the injection point to Lockport Lock and Dam for the 15%, 50% and 90% flow durations is 14, 8 and 21 h, respectively. The Commonwealth Edison power plant plays an important role in the canal flow pattern and CO_2 mixing. As water enters the plant intake creates a stagnation zone between the intake and outfall. On the other hand, water at high velocity from the outfall favours vertical and lateral mixing in the canal downstream.

7.3.3. Dissolved CO_2 distribution with a pulse gas injection. Figure 12 shows the dissolved CO_2 concentration in the riverbed for a pulse gas system at 160 min after the infusion system operation begins. Vertical slices at 10 m and 1000 m from the sparger show the velocity distribution in the canal. Figure 13 show the average CO_2 concentration in these vertical slices. Water with elevated CO_2 moves faster at the centre of the canal where the velocity is the highest. Patches of CO_2 concentration, varying constantly along the canal due to the temporal changes of the gas injection, create a signal contrast that is expected to enhance fish response. The elevated CO_2 gradient close to the diffuser is attenuated as it moves downstream by secondary currents, turbulence mixing, mass transfer at the free surface, and, on a small scale by molecular diffusion. However, the signal-to-background contrast at 1 km from the diffuser is still larger than 60 ppm.

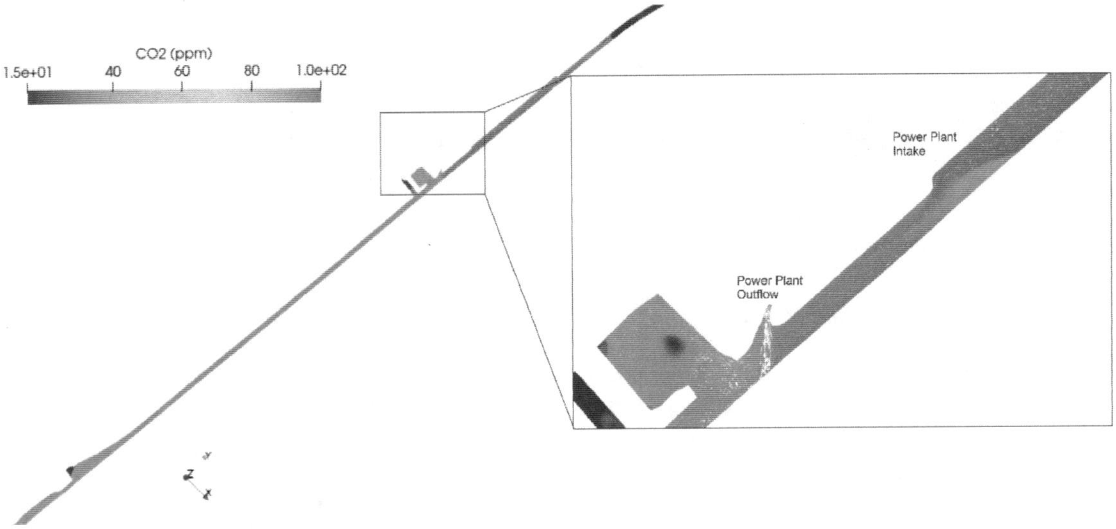

Figure 11. CO_2 concentration and velocity vectors for Sim3 at 1.5 m from the free surface.

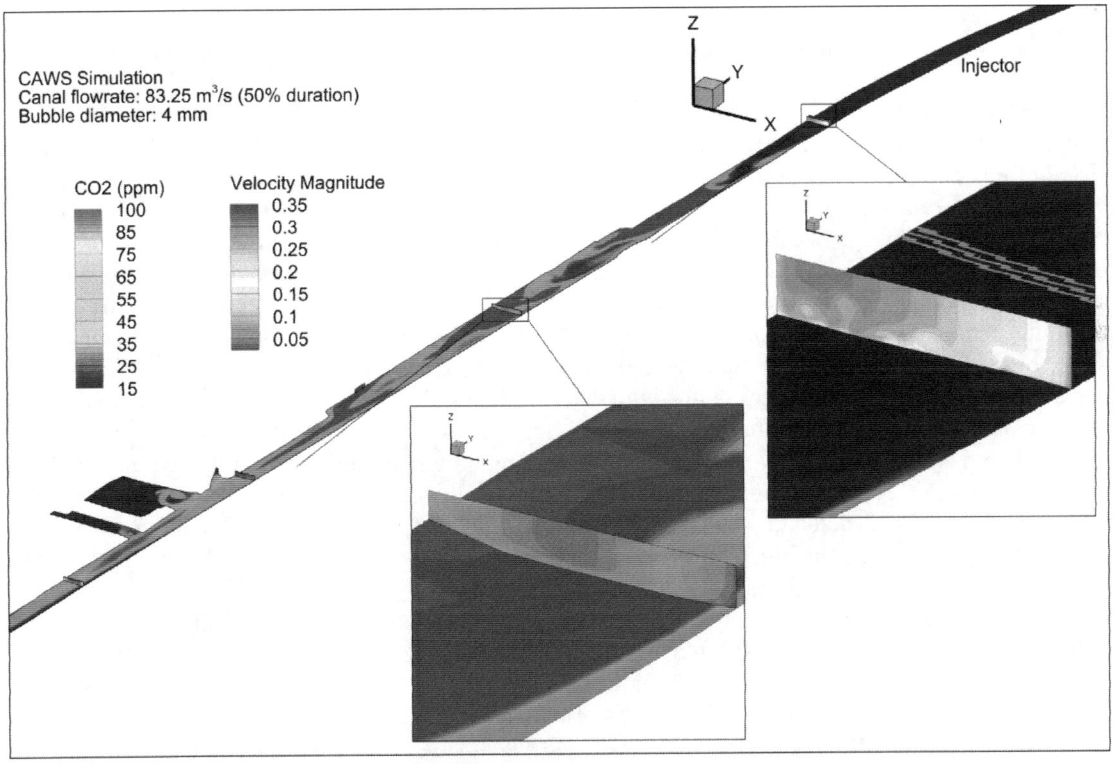

Figure 12. CO_2 concentration for Sim3 at 1.5 m from the free surface and vertical slices with velocity distribution.

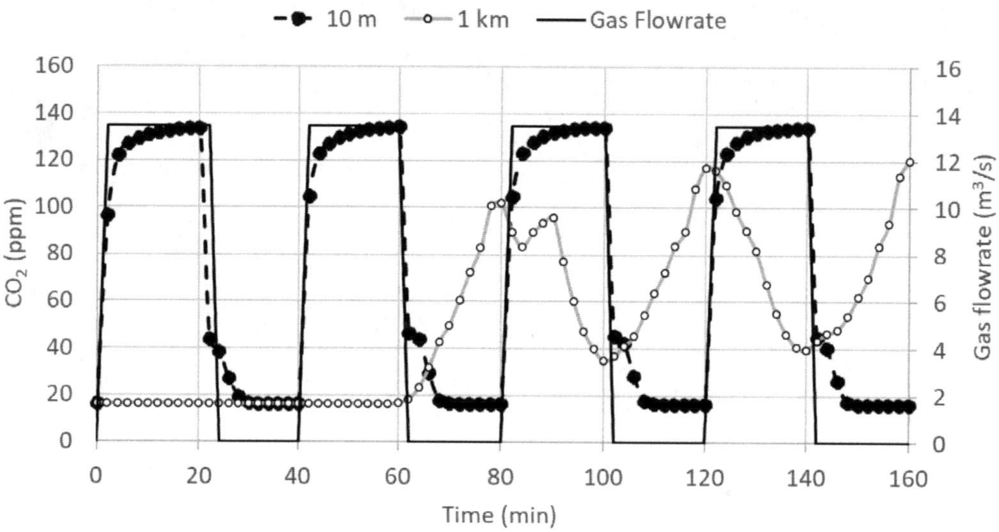

Figure 13. CO_2 concentration and velocity vectors for Sim3 at 1.5 m from the free surface.

8. Conclusions

A simplified two-fluid model was implemented in the OpenFOAM solver buoyantBoussinesqPimpleFoam to assess the feasibility of a CO_2 infusion system to deter fish during the maintenance of the electric barrier in the Chicago Sanitary Ship Canal. The model captures two flow properties, important for the prediction of dissolved CO_2, are captured by the model. The first is the shape of the variable-size bubble plume. The dissolution of bubbles is a function of the local pressure and bubble interfacial area. Consequently, accurate predictions of dissolved CO_2 require knowledge of both pressure and bubble size as the bubbles rise. The second characteristic is the effect of the bubbles on the flow pattern. This is crucial as the large-scale recirculations created by the rising bubbles can synergistically improve fish deterrence.

A constant gas infusion system was numerically evaluated at three canal flowrates and two bubble sizes. Numerical results indicate that the CO_2 gas infusion system produces a deterrence area in the barrier near field region. However, for all simulated conditions, the CO_2 concentration is not fully mixed, creating passageways that fish could use to migrate upstream. Regions with CO_2 concentrations smaller than 100 ppm are found for all simulation conditions. Small bubbles dissolve rapidly near the injection point and thus CO_2 concentration increases with depth. On the other hand, buoyancy of large bubbles transports water with elevated CO_2 concentration toward the free surface. The effect of the small bubbles on the flow pattern is negligible. The injection of 4 mm bubbles results in vertical flow circulations, which are stronger as the canal flowrate decreases.

A CO_2 pulse system with the large bubbles was evaluated with the numerical model. The system creates a CO_2 gradient that is expected to direct fish downstream, thus being more efficient as a deterrence while prolonging the operation time and enhancing the reliability of the setup.

Future work will focus on optimizing gas injection strategies, which requires a comprehensive understanding not only of CO_2 signals that trigger fish behavioural responses but also the dynamics of bubble-induced flows and their deterrence effect on invasive species.

Data Availability Statement

The data that support the findings of this study, including meshes, comparison against experimental data, model setup, and calculation results, are available from the corresponding author upon reasonable request.

Acknowledges
We would like to express our sincere gratitude to the organizing committee of the 18th OpenFOAM Workshop for their support.

References

[1] Ricciardi A. 2006. *Divers. Distrib.* **12** 25.
[2] Patel AG, Glassner-Shwayder K and Eder T. 2010. *Environ. Pract.* **12** 342.
[3] Mayer C, Robinson K and Dettmers J. 2021. *J. Great Lakes Research* **1**, 1.
[4] Rasmussen JL, Regier HA, Sparks RE and Taylors WW. 2011. *J. Great Lakes Research* **37** 588.
[5] Holliman FM, Killgore KJ amd Shea CB. 2015. *ERDC Knowledge Core.* ERDC/TN ANSRP-15-1
[6] Parker AD, Glover DC, Finney ST, Rogers BP, Stewart JG and Simmonds R L Jr.. 2016. Can. J. Fisheries Aquatic Sci **73** 1060.
[7] Kates D, Dennis C, Noatch MR and Suski CD. 2012. *Can. J. Fisheries Aquatic Sci.* **69** 1748.
[8] Nilsson G, Dixson D, Domenici P, McCormick M, Watson S and Munday P. 2012. *Nature Climate Change* **2** 201.
[9] Cupp AR, Erickson RA, Fredricks KT, Swyers NM, Hatton TW and Amberg JJ. 2017. *Can. J. Fisheries Aquatic Sci* **74** 790.
[10] Nilsson G, Dixson D, Domenici P, McCormick M, Watson S and Munday P. 2020. *Nature Climate Change* **1** 790.
[11] Suski CD. 2020. *Fishes* **1**.
[12] Bzonek PA and Mandrak NE. 2022. *Conservation Physiology* **10** 1.
[13] Jasak H. 2009. *Int. J. Naval Architecture and Ocean Engineering* **1** 89.
[14] Zhu S, Motta D, Jackson PR and Garcia MH. 2017. *Environmental Fluid Mechanics* **17** 1.
[15] Deb Roy T, Majumdar AK and Spalding DB. 1978. *Applied Mathematical Modelling* **2** 146.
[16] Grevet JH, Szekely J and El-Kaddah N. 1982. *Int. Journal of Heat Mass Transfer* **25** 487.
[17] Salcudean ME, Lai KYM and Guthrie RIL. 1985. *Can. Journal of Chemical Engineering* **63** 51.
[18] Martinelli L, Talvy S, Liegeois S, Vanden Berghe A, Chauveheid E and Haut B. 2011. *Chem. Engineering Sc.* **66**, 3369.
[19] Castillejos AH and Brimacombe JK. 1987. *Metallurgical Transactions B* **18B**, 659.
[20] Buwa VV, Deo DS and Ranade VV. 2006. *Int. J. Multiphase Flow* **32** 864.
[21] Sheng YY and Irons GA. 1995. *Metallurgical and Material Transactions B* **26B** 625.
[22] Delnoij E, Lammers FA, Kuipers JAM and van Swaaij WPM. 1997. *Chem. Engineering Sc.* **52** 1429.
[23] Fraga B, Stoesser T, Lai CCK and Socolofsky SA. 2016. *Ocean Modelling* **97** 27.
[24] Lain S, Broder D, Sommerfeld M and Goz MF. 2002. *Int. J. Multiphase Flow* **28** 1381.
[25] Castello-Branco MASC and Schwerdtfeger K. 1994. *Metallurgical and Material Transactions B* **25B** 359.
[26] Ishii M. 1975. *Thermo-fluid Dynamic Theory of Two-phase Flow*, 2nd ed. Springer.
[27] Bernard RS, Maier RS and Falvey HT. 2000. *Applied Mathematical Modelling* **24** 215.
[28] Martinelli L, Talvy S, Liegeois S, Vanden BA, Chauveheid E and Haut B. 2011. *Chem. Engineering Sc.* **66** 3369.
[29] Buscaglia G, Bombardelli D and Garcia MH. 2002. *Int. J. Multiphase Flow* **17** 1763.
[30] Antal SP, Lahey RT Jr. and Flaherty JE. 1991. *Int. J. Multiphase Flow* **17** 635.
[31] Lopez de Bertodano M, Lahey RT Jr. and Jones OC. 1994. *Int. J. Multiphase Flow* **20**, 805.
[32] Ishii M and Zuber N. 1979. *AIChE J.* **25** 843.
[33] Politano M, Carrica PM and Converti J. 2003. *Int. J. Multiphase Flow* **29** 1153.
[34] Carrica PM, Bonetto FJ, Drew DA and Lahey RT Jr. 1998. *Int. J. Num. Methods in Fluids* **28** 571.
[35] Takemura F and Yabe A. 1998. *Chem. Eng. Sc.* **53** 2691.

[36] Sato Y, Sadatomi M, Sekoguchi K. 1981. *Int. J. Multiphase Flow* **7** 167
[37] Issa R. 1986. *Journal of Computational Physics* **62** 40.
[38] Versteeg K and Malalasekera W. 2007. *An introduction to computational fluid dynamics: the finite volume method*, 2nd ed. Pearson Education Ltd.

IOP Conf. Series: Materials Science and Engineering **1312** (2024) 012004 doi:10.1088/1757-899X/1312/1/012004

1D and 2D porous media fixed bed reactor simulations with DUO: Steam Methane Reforming (SMR) validation test

E. Daymo[1], A. Tonkovich[1], M. Hettel[2], A. Shirsath[2]

[1] Tonkomo, LLC, Gilbert, Arizona 85297, United States of America

[2] Institute for Chemical Technology and Polymer Chemistry, Karlsruhe Institute of Technology (KIT), Kaiserstr. 12, 76128 Karlsruhe, Germany

eadaymo@tonkomo.com

Abstract. Porous media models are commonly used to simulate packed bed reactors in the chemical process industry. There are several porous media energy equation models available, but commonly a single energy equation for both the fluid and solid phases is selected. Such an approach has been applied to OpenFOAM, with validation testing done using DUO (DETCHEM und OpenFOAM) for a Steam Methane Reforming (SMR) application. Using a popular kinetics mechanism for this reaction, a good fit between simulation and experiment was observed.

1. Introduction

Fixed (aka packed) bed reactors are commonly used in the chemical process industry. The most comprehensive method to analyze fixed bed reactors is to model each individual particle in an approach called Particle Resolved CFD (PRCFD). PRCFD can be highly accurate, but the computational requirements can be too great for many industrial applications. For example, [1] discussed using DUO (DETCHEM [2] und OpenFOAM [3]) to model 86 discrete particles in a packed bed configuration for Catalytic Partial Oxidation (CPOX) and Dry Reforming of Methane (DRM). A typical calculation in [1] required ~8000-core hours without chemical acceleration (ISAT and/or cell agglomeration), and 200 to 800-core hours with chemical acceleration. The DUO solver mentioned above extends OpenFOAM in several ways, including the implementation of surface reactions using a shared library containing the DETCHEM surface reaction solver. DUO is available for many recent versions of OpenFOAM, such as v11, v10, and v2312 (all DUO calculations presented here were done with OpenFOAM v10).

2. Porous media model equations

With respect to the modeling of fixed bed reactors, one way to reduce the computational time (relative to PRCFD) is to use a porous media model. OpenFOAM can model heat transfer in porous media via separate fluid and solid energy equations, with a constant-value fluid-solid heat transfer coefficient to capture the effect of heat transfer between the solid and fluid phases. Also, in the built-in OpenFOAM approach, chemical reactions are evaluated at the fluid temperature. For heterogeneous reactions, which occur at the fluid-solid interface, these model assumptions may not be appropriate, and a single energy equation may be more accurate. As such, the energy equation in DUO (Equation 1) is extended to utilize an effective thermal diffusivity so that the mixed phase (fluid + solid) energy equation is described at steady state by:

$$\nabla \cdot (\rho \cdot u \cdot H_{s,t}) - \nabla \cdot (\alpha_{eff} \cdot \nabla H_s) = -\sum_i^n H_i^o \cdot R_i - \nabla \cdot \sum_i^n H_i \cdot \left((\vec{J_i} + \rho \cdot Y_i \cdot u_c) + \alpha_{eff} \cdot \nabla Y_i \right) . \tag{1}$$

Here n is the number of gas phase species, Y_i is mass fraction, R_i is the source term for the ith specie, ρ is density, u is the superficial fluid velocity, H_s is the sensible enthalpy, $H_{s,t}$ is the total sensible enthalpy ($H_{s,t} = H_s + \frac{1}{2} u \cdot u_c$), H_i^o is the enthalpy of formation (the term $\sum_i^n H_i^o \cdot R_i$ represents the heat of reaction), and H_i is the sensible enthalpy of the ith specie. u_c is the correction velocity, which ensures overall mass conservation ($\vec{J_i}$ is the mass diffusion flux for the ith specie). It is noted that a steady state energy equation is presented here because enthalpy in the time derivative should be calculated using the effective heat capacity, which is not included in the OpenFOAM implementation used for the present work.

α_{eff} is the effective thermal diffusivity (i.e., with units of m^2/s), where $\alpha_{eff} = \frac{k_{eff}}{C_{p_{eff}}}$. $C_{p_{eff}}$ (Equation 2) is the mass weighted effective heat capacity of the porous media (Cp_f and ρ_f are the fluid heat capacity and density, respectively, while Cp_s and ρ_s are the solid heat capacity and density, respectively). The effective thermal conductivity, k_{eff} (Equation 3 [4]), considers both the effective radial fluid, k_{rf}, and effective radial solid, k_{rs}, thermal conductivities. Literature correlations are used for k_{rs} (e.g., Equation 4 from [5]) and k_{rf} (e.g., Equation 5 from [6]), while the thermal conductivity of the fluid mixture, k_f, is calculated from kinetic theory and supplied to DUO by DETCHEM.

$$C_{p_{eff}} = \frac{C_{p_f} \rho_f \, \varepsilon + (1-\varepsilon) C_{p_s} \rho_s}{\rho_f \, \varepsilon + (1-\varepsilon) \, \rho_s} . \tag{2}$$

$$k_{eff} = k_{rs} + k_{rf} \tag{3}$$

$$k_{rs} = k_f \left((1 - \sqrt{1-\epsilon}) + \sqrt{1-\epsilon} \left\{ \frac{2}{1 - \frac{k_f}{k_s} B} \left[\frac{\left(1 - \frac{k_f}{k_s}\right) B}{\left(1 - \frac{k_f}{k_s} B\right)^2} ln \frac{k_s}{k_f B} - \frac{B+1}{2} - \frac{B-1}{1 - \frac{k_f}{k_s} B} \right] \right\} \right) \tag{4}$$

Here $B = C([1-\epsilon]/\epsilon)^{10/9}$, with $C = 1.25$ for spherical particles and 2.5 for cylindrical particles. ε is the bed porosity (volume of voids/total volume), and k_s is the solid thermal conductivity (a constant value supplied by the user).

$$k_{rf} = k_f \left[\frac{\text{Re Pr}}{8.65\left(1 + 19.4\left(d_p/d_t\right)^2\right)} \right] \tag{5}$$

In Equation 5, Re is the particle Reynolds number ($\text{Re} = \rho u d_p/\mu$), Pr is the Prandtl number, d_p is the particle diameter, and d_t is the tube diameter.

An example *heRhoThermoPorous* model (OpenFOAM 10) with the effective thermal conductivity correlations (Equations 2-5) calculated for a user-specified cell zone is provided at https://github.com/TonkomoLLC/OFW18.

3. Steam Methane Reforming (SMR) validation case

To validate the porous media model, DUO results are compared against Steam Methane Reforming (SMR) experimental data presented in [7]. The relevant chemical reactions (1-3) are Steam Reforming of Methane I (chemical reaction 1, Equation 6), Water Gas Shift (chemical reaction 2, Equation 7), and Steam Reforming of Methane II (chemical reaction 3, Equation 8).

$$CH_4 + H_2O \overset{1}{\leftrightarrow} CO + 3H_2 \tag{6}$$

$$CO + H_2O \overset{2}{\leftrightarrow} CO_2 + H_2 \tag{7}$$

$$CH_4 + 2H_2O \overset{3}{\leftrightarrow} CO_2 + 4H_2 \tag{8}$$

The rate equations for chemical reactions 1-3 summarized by [7] (and originally from [8]) are:

$$R_1 = \frac{k_1}{P_{H2}^{2.5}}\left(P_{CH_4}P_{H_2O} - \frac{P_{H_2}^3 P_{CO}}{K_{e1}}\right) \cdot \frac{1}{Q_{SMR}^2} \tag{9}$$

$$R_2 = \frac{k_2}{P_{H_2}}\left(P_{CO}P_{H_2O} - \frac{P_{H_2}P_{CO_2}}{K_{e2}}\right) \tag{10}$$

$$R_3 = \frac{k_3}{P_{H2}^{3.5}}\left(P_{CH_4}P_{H_2O}^2 - \frac{P_{H_2}^4 P_{CO_2}}{K_{e3}}\right) \cdot \frac{1}{Q_{SMR}^2} \tag{11}$$

The various terms in Equations 9-11 are given in Equations 12-17. In Equations 9-17, P_i is the partial pressure of specie i, with units of bar, k_j are rate constants for the jth chemical reaction, and K_i are the adsorption equilibrium constants for specie i. K_{ej} is the equilibrium constant for reaction j, EA_j is the activation energy of chemical reaction j, ΔH_i is the heat of adsorption for specie i, and T is temperature in K.

$$Q_{SMR} = 1 + K_{CO}P_{CO} + K_{H_2}P_{H_2} + K_{CH_4}P_{CH_4} + K_{H_2O}P_{H_2O} + \frac{K_{H_2O}P_{H_2O}}{P_{H_2}} \tag{12}$$

$$k_j = k_{0,j} \cdot e^{\left(\frac{-EA_j}{RT}\right)} \tag{13}$$

$$K_i = K_{0,i} \cdot e^{\left(\frac{\Delta H_i}{RT}\right)} \tag{14}$$

$$K_{e_1} = e^{\left(\frac{-26830}{T}+30.114\right)} \tag{15}$$

$$K_{e_2} = e^{\left(\frac{4400}{T}-4.036\right)} \tag{16}$$

$$K_{e_3} = K_{e_1}K_{e_2} \tag{17}$$

It is noted that due to the presence of P_{H_2} in the denominator of Equation 12, some small amount of H_2 is needed in the feed to prevent division by zero (usually a small initial mass fraction of 10^{-4} to 10^{-5} H_2 is sufficient to avoid this numerical issue). Also, the equilibrium expression for chemical reaction 1 (Equation 15) is from [7], but the equilibrium equation for chemical reaction 2 (Equation 16) is from [9]. [7]'s equilibrium expression for Water Gas Shift is not used since [7] derived its equilibrium equation for this reaction from its own experimental data (which were collected in a narrow temperature, pressure and composition window). The equilibrium expression for the Water Gas Shift reaction from [9] is applicable to broader operating conditions. Also, since chemical reaction 3 is the sum of chemical reactions 1 and 2, the equilibrium expression for chemical reaction 3 is the product of the equilibrium constants of the first two chemical reactions (Equation 17). The rate expressions (Equations 9-11) have net overall units of mol/kg cat/s, and the kinetic model parameters are summarized in Table 1. Although the kinetic expressions are complex, the kinetic expressions can be entered easily using DETCHEM's User Defined Function (UDF) capability.

Parameter	**Value**	**Units**
$k_{0,1}$	9.048x10^{11}	bar$^{0.5}$
EA_1	209,500	J/mol
$k_{0,2}$	5.43x10^5	Bar^{-1}
EA_2	70,200	J/mol
$k_{0,3}$	2.14x10^9	bar$^{0.5}$
EA_3	211,500	J/mol
$K_{0,CO}$	8.11x10^{-5}	bar^{-1}
ΔH_{CO}	-70,230	J/mol
K_{0,H_2}	1.68E4	bar^{-1}
ΔH_{H_2}	85,770	J/mol
K_{0,CH_4}	1.995x10^{-3}	Pa^{-1}
ΔH_{CH_4}	-36,650	J/mol
K_{0,H_2O}	1.68x10^4	-
ΔH_{H_2O}	85,770	J/mol

Table 1. Kinetic parameters summarized by [7] and used in the SMR validation test.

Reactor design parameters common to all cases are summarized in Table 3, the test matrix is listed in Table 4, and additional DUO (i.e., OpenFOAM) boundary condition information is provided in Table 5. It is noted in Table 5 that the external wall temperature boundary condition is *externalWallHeatFluxTemperature*. The wall temperature (i.e., the "Ta" field in the OpenFOAM boundary condition) is set to the value listed in Table 3 (which changes from case-to-case). The heat transfer coefficient ("h" in OpenFOAM) is set to a constant value of 1,700 W/m^2/K. This constant was estimated with the 1D packed reactor solver DETCHEM^{PBR}, but it is observed for these cases that the results are not highly sensitive to the value of the external heat transfer coefficient. A constant temperature boundary condition is not appropriate, because as discussed by [10], when the combined fluid and solid energy equation approach is used for the porous media model (also called the α_w model), the wall boundary should be set to a heat flux condition. The OpenFOAM grid is a 2D wedge with 35,500 cells, created with blockMesh and makeAxialMesh (https://github.com/ZhangYanTJU/makeAxialMesh).

As shown in Figure 1, a parity plot (calculated CH_4 conversion vs. experimental CH_4 conversion) demonstrates an excellent match between experimental data and the DUO model ($R^2 = 0.9843$). Due to the endothermic nature of Steam Methane Reforming, the temperature gradients can be large (e.g., a ~250°C temperature gradient is observed for one of the cases in Figure 2).

Parameter	**Value**	**Units**
Reactor length	0.15	m
Reactor diameter	1	cm
Particle diameter	1.75	mm
Porosity	0.4	-
k_s	1	W/m/K
d_t	10	mm
P_{outlet}	1.5	bar (absolute)

Table 2. Reactor design parameters from [7] for the SMR validation test.

Case	T_{inlet} & T_{wall} (°C)	H_2O:CH_4 feed ratio	Inlet velocity (m/s)
1	800	3	7.478
2	700	3	6.781
3	600	3	6.084
4	500	3	5.387
5	800	3	8.948
6	700	3	8.114
7	600	3	7.281
8	500	3	6.447
9	800	4	9.347
10	700	4	8.476
11	600	4	7.605
12	500	4	6.734

Table 3. Test matrix for the comparison between DUO and experiments (Figure 1).

OpenFOAM Parameter	Inlet boundary condition	Outlet boundary condition	Wall
T	fixedValue, see Table 3	zeroGradient	externalWallHeatFluxTemperature
p	zeroGradient	fixed Value, see Table 2	zeroGradient
U	fixedValue, see Table 3	zeroGradient	no slip
Species mass fraction	fixedValue, see Table 3; H_2 is set to $1x10^{-4}$	zeroGradient	zeroGradient

Table 4. DUO (i.e., OpenFOAM) boundary conditions for the test cases

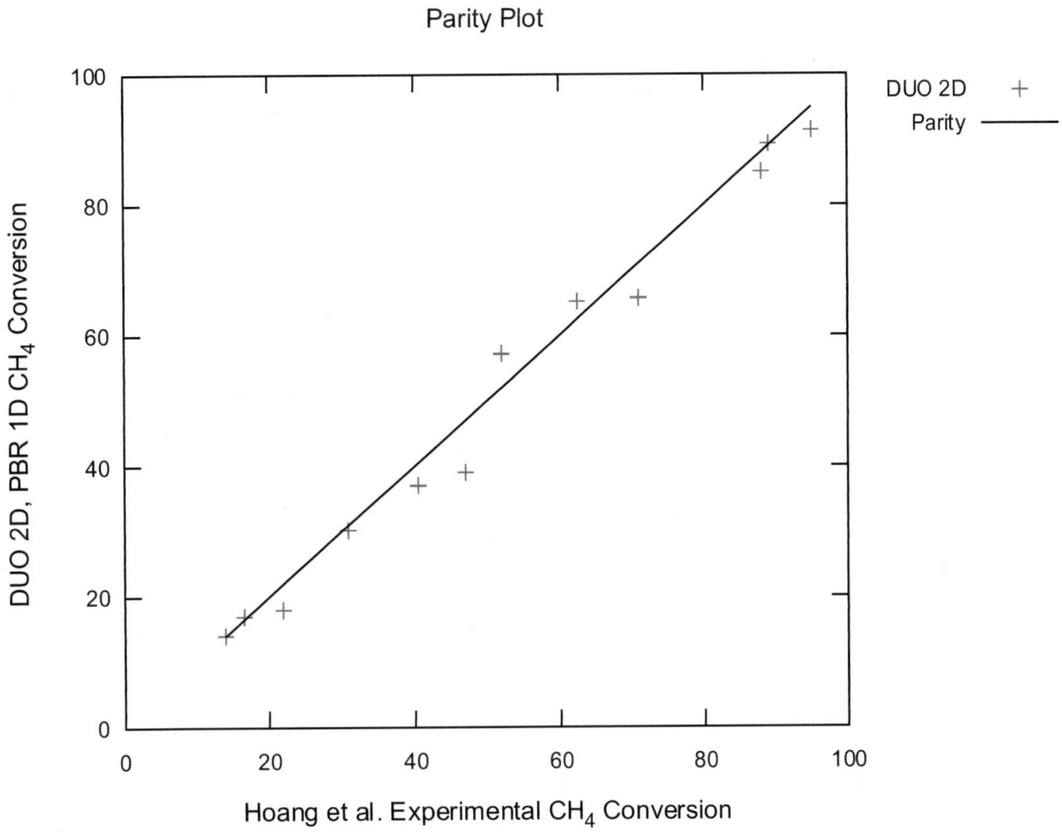

Figure 1. Parity plot of DUO 2D and experimental data for experiments reported by [7].

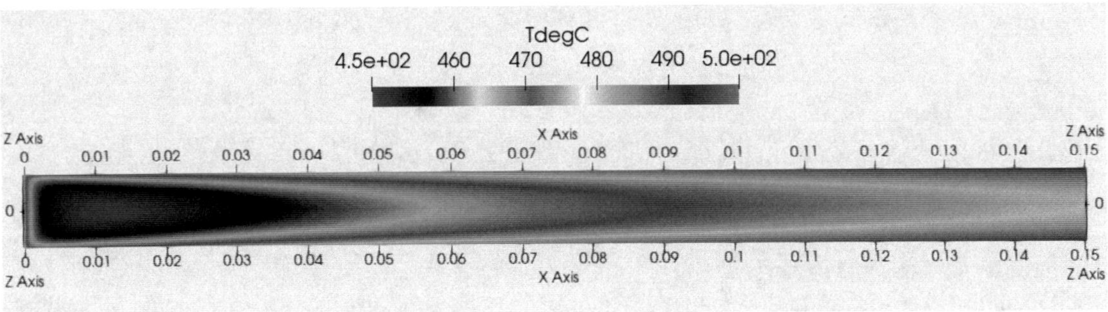

Figure 2. Temperature gradients in the SMR reactor can be significant, such as the ~250°C temperature gradient observed in one of the experimental conditions (Case 9 in Table 2).

4. Conclusions

Porous media modeling is important for the analysis of non-isothermal packed bed reactors, a key unit operation in many chemical processes. OpenFOAM allows for separate fluid and solid energy equations with porous media, with chemical reactions occurring at the fluid temperature. This may not be the best assumption for heterogeneous catalytic reactions, which occur at a solid surface. To address such heterogeneous reactions, the solver DUO was extended to model packed bed heat transfer. DUO was then validated against literature data for the Steam Methane Reforming reaction. The resulting parity plot showed good coincidence between experimental data and DUO calculations, validating the approach.

References

[1] Daymo E, Hettel M, Deutschmann O and Wehinger G 2022 Accelerating particle-resolved CFD simulations of catalytic fixed-bed reactors with DUO *Chemical Engineering Science* **250** 117408 doi: 10.1016/j.ces.2021.117408

[2] Deutschman O, Tischer S, Correa C, Chatterjee D, Kleditzsch S, Janardhanan V M, Mladenov N, Minh H D, Karadeniz H, Hettel M, Menon V, Banerjee A, Gossler H, Shirsath A, and Daymo E 2022 DETCHEM Software Package 2.9 ed. www.detchem.com

[3] OpenFOAM, the Open Source CFD Toolbox 2022 www.openfoam.org

[4] Yagi S and Kunii D 1957 Studies on effective thermal conductivities in packed beds AIChE J **3**(3) 373-381 doi: https://doi.org/10.1002/aic.690030317

[5] Specchia V, Baldi G and Sicardi S 1980 Heat transfer in-packed bed reactors with one phase flow *Chem. Engng Commun* **4** 361-380 doi: https://doi.org/10.1080/00986448008935916

[6] Zehner P and Schluender E U 1970 *Ingr. Tech.* **42** 933 doi: https://doi.org/10.1002/cite.330421408

[7] Hoang D H, Chan S H, Ding O L 2005 Kinetic and modeling study of methane steam reforming over sulfide nickel catalyst on a gamma alumina support *Chemical Engineering Journal* **112** 1-11 doi: https://doi.org/10.1016/j.cej.2005.06.004

[8] Xu J and Froment G F 1989 Methane steam reforming, methanation and water-gas shift: I. Intrinsic kinetics AIChE J **35** 88–96 doi: https://doi.org/10.1002/aic.690350109

[9] Abbas S Z, Dupont V, Mahmud T 2017 Kinetics study and modelling of steam methane reforming process over a NiO/Al_2O_3 catalyst in an adiabatic packed bed reactor *International Journal of Hydrogen Energy* **42** 2889-2903 doi: https://doi.org/10.1016/j.ijhydene.2016.11.093

[10] Stegehake C, Riese J and Grünewald M 2019 Modeling and validating fixed-bed reactors: A state-of-the-art review *ChemBioEng Rev* **6** (2) 28–44 doi: https://doi.org/10.1002/cben.201900002

Acknowledgements

The authors would like to thank the organizing committee of the 18th OpenFOAM Workshop for the support provided. A cost-free academic license of DETCHEM™ by the omegadot software & consulting GmbH is also gratefully acknowledged.

Availability of data and materials

The OpenFOAM software used to model 1D and 2D heat transfer effects in porous media is available at https://github.com/TonkomoLLC/OFW18. Cases (including meshes), comparison data, and calculation results presented in this paper are available from the corresponding author on reasonable request. Inquiries regarding DETCHEM and DUO can be made to omegadot software and consulting GmbH (https://www.omegadot.software).

OpenFOAM simulations for development of a mist flow diverter in aerosol jet printing

James Q Feng

Optomec Inc., 2575 University Avenue, #135, St Paul, Minnesota, USA

Email: james.q.feng@gmail.com

Abstract. To design functional devices meeting the performance requirements for Aerosol Jet® (AJ) printing to deposit ink on substrate by a collimated impinging mist jet, understanding the internal mist flow behaviour is important. Herewith a computational fluid dynamics (CFD) model is presented with an OpenFOAM software package, for development of a mist flow diverter in the AJ printing equipment. The function of such a mist flow diverter is to direct a mist flow with specified flow rate to the deposition nozzle for printing the ink material onto substrate and to divert the mist flow away when the nozzle moves over a substrate location where ink deposition is unwanted. The CFD results can reveal detailed mist flow behaviour during mist flow switching for quantitative analysis of mist front travelling time through different regions. It is shown that such an internal pneumatic diverter inevitably brings noticeable on-delay and off-delay, more so for finer feature printing with lower mist flow rate due to time needed for mist droplets to travel from the mist flow switching chamber to substrate. Moreover, the on-delay and off-delay also come with gradual fade-in and fade-out effects in printed features especially when the substrate is moving at a considerable speed. Presented here is but one example of CFD analysis applied to performance examination of a commercial vacuum-boost diverter, although the established computational model has also been utilized to guide equipment design with evaluations of the effects of different geometric shapes and sizes by virtue of the convenience for CFD domain geometry and mesh adjustments.

1. Introduction

The Aerosol Jet® (AJ) direct-write technology enables material deposition for producing micron-scale fine features on various substrates with almost arbitrary geometry, which is of particular interest in manufacturing 3D conformal electronics among other applications. It operates by delivering an aerosol mist stream consisting of micron-size "ink" droplets, through an aerodynamic focusing nozzle with co-flowing sheath gas in a tapered channel, to the substrate several millimetres away with a collimated impinging jet flow [1]. The pattern of ink material produced with AJ printing is essentially controlled in a "vector drawing mode" according to programmed "toolpath" with specified relative motion of the substrate relative to the deposition nozzle. In effect, the toolpath consisting of instructions for substrate moving velocity from location to location maps the substrate coordinate positions to a corresponding time series. When a constant mist flow rate is fed into the deposition head, a mechanism for switching the mist flow away from substrate must be implemented to prevent ink deposition to where ink is unwanted along the moving toolpath. In other words, the mist stream toward substrate needs to be turned off at the time when the places on substrate where ink deposition

IOP Conf. Series: Materials Science and Engineering **1312** (2024) 012005 doi:10.1088/1757-899X/1312/1/012005

is unwanted are moved under the AJ deposition nozzle, and to be turned on at the time when ink deposition is wanted. For the mist flow toward substrate, it can be considered as a function of time with on or off at certain time points.

For many years, an earlier version of AJ mist shuttering device called external mechanical shutter had been implemented in the AJ systems, by electromechanically moving a small blade in and out of mist flow path between the AJ deposition nozzle and substrate (external of the deposition head). It serves the purpose of turning on and off mist deposition, while being limited to short printing time due to the fact that ink on the small shuttering blade, with restricted ink holding volume, could spill over. It also necessarily takes more than a millimetre space off the few millimetres standoff—valuable for printing over substrates with complicated geometries—between the nozzle tip and substrate. To remove those shortcomings of the external mechanical shutter, a pneumatic mist diverter inside the deposition head has been developed by switching the mist flow to and away from the deposition nozzle inside the AJ deposition head. In the present work, a vacuum-boost mist flow diverter is investigated for rapid mist flow on-and-off switching in AJ printing. Like recent development of several other AJ print-head components, the OpenFOAM CFD simulations have played important roles in providing enlightening insights to the mist flow behaviour difficult to visualize in typical laboratory settings [2-4]. Here also, OpenFOAM simulations were used to aid the design and development of this vacuum-boost mist flow diverter (as a commercial AJ device) by enhancing fundamental understanding of its operational behaviour.

Because the distance between the mist flow switching location in an internal pneumatic diverter is much longer than that of mechanical shutter, longer time is needed for the mist front to travel from the mist switching location to substrate resulting in a more noticeable time delays to be factored in the toolpath programming with the shutter points. Hence a quantitative evaluation of shutter delays with OpenFOAM simulations becomes the main focus of the present work, although the established computational model has also been utilized for the equipment designs by evaluating and optimizing the effects of different geometric shapes and sizes in various prototypes, by virtue of the convenience for domain geometry and mesh adjustments.

2. Computational model description

As the mist flow in AJ systems being usually considered as a two-phase particle-laden flow with the ink micro-droplets of 1 to 5 microns treated as spherical particles [3, 4], the mist flow behaviour in a vacuum-boost diverter can be reasonably simulated with a Lagrangian solver implemented by adding a *kinematicCloud* class to the *pimpleFoam* solver for a transient incompressible laminar flow based on the OpenFOAM-6 package [5]. It involves a two-way coupling between the *kinematicCloud* parcels and carrier fluid flow. To account for the non-continuum effects on small particle drag force, the source code of SphereDragForce.C (in src/lagrangian/intermediate/submodels/Kinematic/ ParticleForces/Drag/SphereDrag) is modified by dividing the traditionally computed drag force *value.Sp()* for a particle of diameter d_p by C_c where

$$C_c = 1 + \frac{\lambda}{d_p}\left[A_1 + A_2 \exp\left(-A_3 \frac{d_p}{\lambda}\right)\right] \quad (1)$$

denotes the Cunningham slip correction factor with $\lambda = 0.065$ μm (at 25 °C), $A_1 = 2.514$, $A_2 = 0.8$, and $A_3 = 0.55$ [6] as was previously done by Feng [7, 8] in other applications of OpenFOAM Lagrangian solvers.

Shown in figure 1 is the computational domain with mesh generated with the OpenFOAM meshing utility *blockMesh* with the GNU preprocessor m4 for convenient parametric mesh adjustment, consisting of a mist inlet tube connected to a vacuum chamber, a middle tube between the vacuum chamber and the sheath chamber, and the nozzle channel tube leading to the deposition nozzle. At the upstream end of the inlet tube is the mist flow inlet, and the sheath gas inlet is located around the circular side of the sheath chamber. On the top of the vacuum chamber is the vacuum flow outlet, and a flow outlet to the aerodynamic focusing nozzle (which is the same as the nozzle inlet) is

IOP Conf. Series: Materials Science and Engineering **1312** (2024) 012005 doi:10.1088/1757-899X/1312/1/012005

located at the downstream end of the nozzle channel tube. The rest of boundaries are all assigned as solid walls.

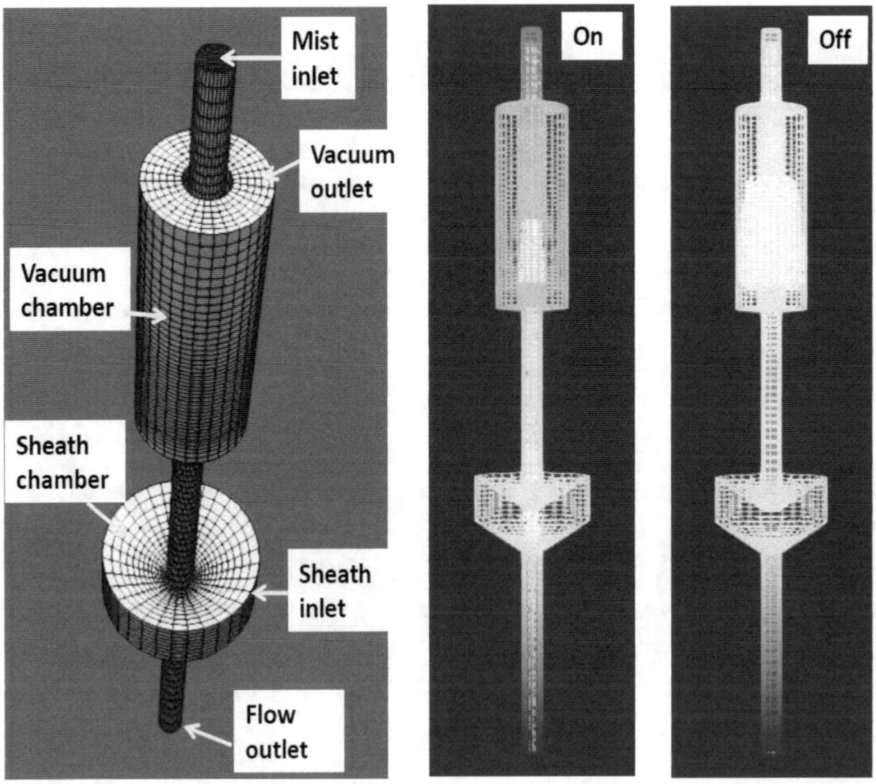

Figure 1. Concentric computational domain for mist flow simulations in a vacuum-boost diverter, and snapshots of paraFoam display of mist flow patterns when mist flow is switched on (mist flows through outlet) and off (no mist toward outlet) where the pressure-colour code (blue for $p = 0$ and red for $p = 6.9$ Pa) indicates that pressure in the sheath inlet chamber remains constant (about $p = 4.5$ Pa) during diverting.

For the carrier fluid flow, the stratighforward no-slip and no-penetration boundary conditions are applied at the solid walls . The *flowRateInletVelocity* type condition is applied with a fixed flow rate Q_m specified for U at the mist inlet but with table files of time-dependent flow rate for the volumetricFlowRate specified for U at the sheath inlet and vacuum outlet, respectively. The boundary condition for p is zeroGradient for all boundaries except a fixedValue of 0 is specified at the flow outlet where the *zeroGradient* type of boundary condition is used for U. The *kinematicCloud* parcels are introduced with an OpenFOAM® build-in particle injection model called *coneNozzleInjection* type for a disk particle source placed close to the mist inlet, as specified in *kinematicCloudProperties* dictionary. The volume percent of randomly injected particles is assumed to have a *RosinRammler* type of diameter distribution with *minValue* = 3e-07, *maxValue* = 1.5e-05, d = 2.7e-06, and n = 2.4 corresponding to the ink droplets of diameters between 1 to 5 microns for AJ printing [1]. Also in *kinematicCloudProperties* the *localInterationCoeffs* are assigned as type stick on walls as well as on outlets while type rebound with e = 0.97 and mu = 0.09 on inlets for the *patchInteractionModel*. For the carrier gas (such as the nitrogen typically used in AJ printing), the density is specified as 1.2 kg m^{-3} and kinematic viscosity 1.5e-05 m^2 s^{-1} in the *transportProperties* dictionary.

3. Results and discussion

For the particular configuration of a vacuum-boost diverter under consideration, the "on" snapshot in figure 1 shows that the mist is flowing toward deposition nozzle (through the flow outlet), driven by a pressure gradient with high (red) at the mist inlet and low (blue) at the flow outlet, in the absence of the vacuum flow. The "off" snapshot shows no mist flow toward the flow outlet to deposition nozzle, when the vacuum flow is turned on creating a relatively lower pressure in the vacuum chamber with respect to that in the sheath chamber such that the mist flow is diverted upward toward the vacuum flow outlet. To minimize the on and off delays, it is desirable to maintain pressure in the sheath chamber constant [9]. This can be accomplished by introducing a boost flow into the sheath chamber with a flow rate matching the vacuum flow rate during diverting. For example, the nominal case shown in figure 1 has a mist flow of Q_m = 30ccm and sheath flow rate of Q_{sh} = 60ccm for the "on" situation. For the "off" situation, we have an additional vacuum flow of Q_v = 45ccm (= 1.5 x Q_m) and simultaneously a boost flow of $Q_b = Q_v$ = 45ccm added to the sheath inlet; so the total flow rate at the sheath inlet now is $Q_{sh} + Q_b$ = 105ccm. In other words, there is a portion of Q_b (~15ccm) from the sheath chamber to vacuum chamber, carrying the mist in the middle tube toward vacuum flow outlet, while the total flow rate toward the flow outlet remains the same as $Q_{sh} + Q_m$ = 90ccm with pressure in the sheath chamber remains constant as visually indicated from the pressure-colour code in figure 1.

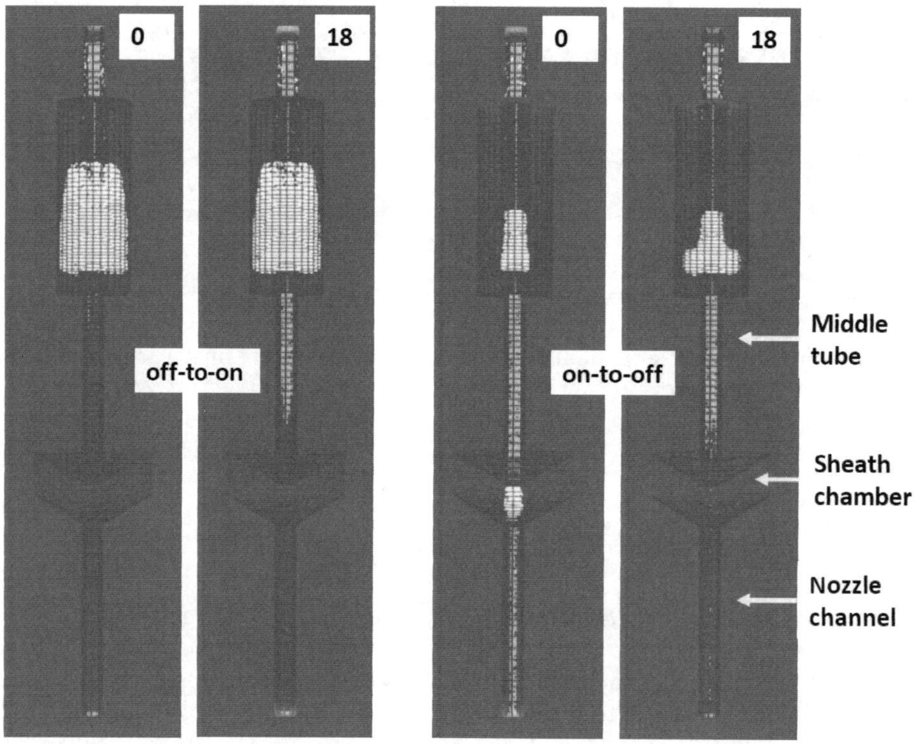

Figure 2. Snapshots of paraFoam display of mist flow patterns in the nominal case after off-to-on switching and on-to-off switching at t = 0 and t = 18ms.

The results of transient CFD simulations can be examined when visualized with a postprocessing tool called paraFoam (based on ParaView) for further insights into the behaviour of this vacuum-boost diverter. Figure 2 shows the mist front moving downward in the middle tube, from the vacuum chamber to sheath chamber, with a parabolic-like profile, after 18ms from switching off-to-on. Given enough time, the mist front will eventually move through the nozzle channel outlet (as seen in the initial image for the on-to-off situation) into the deposition nozzle, for aerodynamic focusing to reach the substrate as a collimated mist jet stream (as shown in Figure 3 which displays images of actual commercial AJ equipment, deposition head, etc.) Clearly, the co-flowing sheath gas entering from the sheath chamber prevents direct contact of mist droplets with the nozzle channel wall, besides its function for aerodynamic focusing in the deposition nozzle. Figure 2 also shows the situation of switching on-to-off where mist is still moving upward in the middle tube while only a few droplets still left in the nozzle channel after 18ms from the time of switching. These residual droplets will reach the substrate resulting in a so-called fade-out tail in printed feature if the substrate is moving at a considerable speed. When switching from on to off, it takes a longer time for the cleaning all the mist droplets in the middle tube than the mist front travel time because of these droplets close to the tube wall with very low velocity.

Table 1 presents the mist travel time in the middle tube (14mm long), sheath chamber gap (2.8mm) between the middle tube outlet and inlet of the nozzle channel, and nozzle channel (12.5mm long), as well as "On-delay" and "Off-delay" for various mist flow rate and corresponding sheath flow rate. Both the middle tube and nozzle channel have the same diameter of D = 1.4mm. The value in parentheses for "Off-delay" includes the time needed to clean out almost all the droplets in the middle tube. While the "On-delay" is simply $t_m + t_g + t_{ch}$, the "Off-delay" is always longer than $t_g + t_{ch}$ due to the extra time needed for the boost flow adjustment to pinch off mist stream in the sheath chamber. During on-to-off switching, the mist stream is pinched off inside the sheath chamber with the lower half moving toward the deposition nozzle and upper half toward the vacuum chamber. In the nozzle channel, the mist moves much faster driven by an extra flow rate of the sheath gas.

Table 1. Mist travel time in the middle tube (t_m), sheath *chamber* gap (t_g), and nozzle channel (t_{ch}) as well as On-delay and Off-delay, in units of millisecond, for mist flow rate (Q_m) of 12, 30, 60ccm with corresponding sheath flow rate (Q_{sh}).

Q_m	Q_{sh}	t_m	t_g	t_{ch}	On-delay	Off-delay
12	24	55	13	17	85	45 (90)
30	60	23	4	7	34	22 (77)
60	120	12	2	3	17	11 (60)

The values of t_m and t_{ch} in table 1 are actually quite close to that estimated by using the peak velocity of the Poiseuille tube flow for mist front motion as

$$t = \frac{\pi D^2 L}{8Q} \tag{2}$$

with D and L denoting the tube diameter and length, Q the flow rate. For example, equation (2) would yield t = 54ms for Q = 12ccm and L = 1.4cm and t = 16ms for Q = 36ccm and L = 1.25cm, about the same as t_m and t_{ch} for the Q_m = 12ccm case. A slight difference between (2) and t_m, t_{ch} in table 1 is expected from the entrance effect in tube flow. This effectively verifies the CFD model. However, the insights into detailed behaviour of mist flow in the sheath chamber can only be gained with the CFD simulations.

4. Concluding remarks

The CFD results suggest that the internal pneumatic diverter inevitably brings longer on-delay and off-delay (> 10ms, more for finer feature printing with lower mist flow rate, e.g., Q_m < 30ccm) than that with the external mechanical shutter (< 2ms), due to extra time required for mist droplets to move from the mist flow switching (vacuum) chamber to the substrate. Moreover, the on-delay and off-delay also come with fade-in and fade-out effects in printed features especially noticeable when the substrate is moving at a considerable speed. Such delays are proportionally more significant for finer feature printing limited to lower mist flow rate due to the jet flow stability constraint [1]. In principal, these annoying print quality problems can be completely eliminated by a well-designed toolpath with judicious usage of dynamic variables such as "shutter-in-the fly", "acceleration", "deceleration", "speed", etc. available in the motion control software toolbox. Of course, perfecting a well-designed toolpath takes time and efforts. Yet these efforts seem worthwhile especially for production applications because the toolpath is expected to be repeatedly used for a large quantity of printed parts; whether the toolpath is well-designed can have significant economic impact on process efficiency and product quality.

Figure 3. Images of an actual Optomec commercial Aerosol Jet® printing equipment (Aerosol Jet FLEX System - Optomec), a deposition head with vacuum-boost diverter (AJ FLEX - Optomec), and a collimated mist jet stream from the deposition nozzle to substrate (Optomec Engineer Explores Control of Mist Flow in Aerosol Jet Printing - 3DPrint.com | The Voice of 3D Printing / Additive Manufacturing). Copyright on the images is held by the contributors. Apart from Fair Use, permission must be sought for any other purpose.

On the other hand, the CFD results with visualized motions of mist droplets can also elucidate possible improvements to the vacuum-boost diverter. For example, the on-delay and off-delay can be reduced by shrinking the diameters of nozzle channel and middle tube; a decrease of the channel diameter by a factor of 2 can reduce the mist travel time by a factor of 4 according to (2). But the tube diameter as well as the tube length cannot be reduced arbitrarily, for practical reasons such as the requirements of chamber volume, nozzle form factor, robust functionality, and so on. Moreover, adding a sheath gas to the middle tube mist flow, similar to that in the nozzle channel, can certainly help reduce the on-, off-delays, but it will increase the device cost and complexity often also at the expense of decreased printing throughput. Thus, a complete solution to such print quality problem associated with mist flow delays seems to be investing efforts in perfecting the toolpath design with well-equipped software capabilities. In doing so, the present computational model can provide quantitative prediction of mist flow delays for any possible combinations of mist flow rate Q_m and sheath flow rate Q_{sh} which must be determined on case-by-case basis depending on the printing feature size and ink properties.

Presented here is but one example of CFD analysis applied to performance examination of a commercial equipment (as displayed in Figure 3) with the AJ technology. By virtue of the convenience for domain geometry and mesh adjustments, the established computational model has also been utilized for the equipment designs by evaluatingAlso and optimizing the effects of different geometric shapes and sizes. In this regard, the OpenFOAM (with its highly functional, freely available open source software) has provided substantial opportunities not only for academic research institutions but also for small organizations and companies to conduct CFD simulations relevant to research projects or product design and development. Its complete software package including source code offers the users with great freedoms for code modification to fit specific application requirements. Also comes with each tested functional solver is a set of exemplifying tutorial cases, which can be very helpful to average users who may not immediately have the time for fine-tuning the numerical scheme parameters but want to quickly start computing solutions for the problem at hand, with a default set of parameter values often able to at least keep the computation going. In general, the scientific knowledge and understanding gained with and through the OpenFOAM simulations have significantly enhanced our engineering efforts in both equipment design and process development for AJ applications.

Acknowledgments

The author would like to thank Dr. Mike Renn, Dr. David Sessoms and Justin Bourassa for insightful discussions. Also, initial helps from Dr. Alejandro Lopez at the University of Strathclyde for modifying source code of the OpenFOAM Lagrangian solvers should be greatly appreciated. The publication of the present work is made possible with support provided by the organizing committee of the 18th OpenFOAM Workshop.

References

[1] Feng JQ and Renn MJ 2019 Aerosol Jet® direct-write for microscale additive manufacturing. *Journal of Micro- and Nano-Manufacturing* **7** 011004

[2] Feng JQ 2014 Aerosol Jet® material deposition behavior simulated with the interFoam VOF solver. OFW09.0024, 9th OpenFOAM Workshop, Zagreb, Croatia, 23-26 June

[3] Feng JQ 2015 Aerosol Jet® material deposition behavior simulated with an OpenFOAM Lagrangian solver. OFW10-28, 10th OpenFOAM Workshop, Ann Arbor, USA, 28 June-2 July

[4] Feng JQ 2016 OpenFOAM simulations for Aerosol Jet® technology development. 3rd Midwest OpenFOAM Users' Group Meeting, Madison, USA, 15 October; 2019 4th Midwest OpenFOAM User Group Meeting, Minneapolis, USA, 5 October

[5] OpenFOAM v6 User Guide. https://doc.cfd.direct/openfoam/user-guide-v6/

[6] S. K. Friedlander SK 1977 *Smoke, Dust, and Haze, Fundamentals of Aerosol Behavior* (John Wiley & Sons)

[7] Feng JQ 2017 A computational study of particle deposition patterns from a circular laminar jet. *Journal of Applied Fluid Mechanics* **10(4)** 1001-1012

[8] Feng JQ 2023 Aerosol deposition in 90° circular tube bends with laminar flows: Effects of inertial impaction and gravitational settling. *Aerosol Science and Engineering* **7(1)** 107-117

[9] Christenson KK, Renn MJ, Paulsen JA, Hamre JD, Conroy C and Feng JQ 2020 Shuttering of aerosol streams. *US Patent* 10632746B2

A convergence study simulating regular waves using the k-ω SST turbulence model in OpenFOAM®

Xihang Xu[1*], Md Salauddin[1] and Jennifer Keenahan[1]

[1]School of Civil Engineering, University College Dublin, Dublin 4, Ireland

*xihang.xu@ucdconnect.ie

Abstract. OpenFOAM® has been widely used in coastal engineering to visualise and analyse wave-structure interaction and evaluate the effectiveness of innovative coastal protection structures. To study the influence of an eco-retrofitted seawall on the wave overtopping process, a 2D numerical wave tank with a k-ω SST turbulence model is developed to recreate previously completed experiments. Results of wave structure interactions at a plain vertical seawall subjected to regular waves using OpenFOAM® are presented, considering various configurations of time steps and grid resolutions. Both deep water (near the wave maker) and shallow water (near the structure) wave structure interactions are analysed. Results indicate that a finer mesh density is required in shallow water regions compared with deep water regions due to the dynamics of the wave-breaking process in the wave reflection area. Results also indicate that mesh density has a more significant influence than the time step. Overall, this paper showcases the entire process of establishing two-phase flow CFD (computational fluid dynamics) simulations in OpenFOAM® and provides a reference model for future research on studying the effectiveness of eco-retrofitting approaches on wave attenuation.

1. Introduction

Coastal protections, including seawalls and breakwaters, are widely applied worldwide to meet the challenges of extreme climatic hazards such as wave-induced overtopping and runup [1]-[2]. Evidence is mounting on the increasing threat to functional and structural safety of coastal protections due to more frequent and extreme climatic hazards often associated with global and regional climate change and rising sea levels [3]-[4]. Meanwhile, greater public awareness of environmentally friendly development demands a 'greener' and more sustainable design for coastal structures [5]-[6]. In such a context, an 'Eco-retrofitted approach' is one of the most promising solutions, which incorporates some biomimicry-based interventions to the sea defences, such as attaching 'flowerpots' or 'textured concrete tiles' to the seawall or creating 'drill-cored rock pools' in the breakwaters [7]-[8]. The ecological benefits of these interventions, which increase the surface roughness of attached structures, have been widely reported [9]-[11].

However, only a few studies evaluated the engineering benefits of such interventions, and most of the eco-retrofitted designs are mainly based on the ecological benefits, lacking guidance from an engineering perspective based on the understanding of wave structure interactions [12][14]. The wave overtopping analysis is one of the most notable components in evaluating the engineering performance of artificial sea defences [15]-[17]. Experiments, including scaled or full-size tests, are the traditional methods adopted in generating wave-structure interaction analysis, which is costly, time-consuming and poses challenges in conducting parametric studies.

With recent developments in numerical methods and computational capacity, computational fluid dynamics (CFD) has become a feasible approach for visualizing the phenomenon and conducting detailed parametric studies of complex wave structure interactions [17],[19]. OpenFOAM®, as an open-source CFD toolbox, is one of the most attractive modelling approaches which can simulate two-phase flows with different turbulence models and has been widely applied in industry and academia [20],[23]. Nevertheless, the primary prerequisite for carrying out a comprehensive parametric study by numerical methods is the validation of the developed models because the precision and reliability of the models are highly dependent on the parameter settings such as turbulence model, mesh density and the time steps [24]-[25]. It is well known that the smaller the cell size and time step set, the more accurate the modelling results will be, but this will lead to significant computational time and storage demands.

Turbulent effects are well-considered in the wave analysis, and the accuracy of water elevation simulation is highly dependent on the selection of different RANS turbulence models e.g., $k-\omega$, $k-\varepsilon$ and $k-\omega\ SST$ turbulence model [26]-[27]. Based on the previous studies, it is found that the $k-\omega\ SST$ turbulence model has the best performance in predicting the wave elevation among these models [28].

The dimensionless wall distance (y^+), defined as the normal distance from the first cell to the boundary wall, considerably influences the prediction of the near-wall flow [29]. Therefore, selecting a suitable y^+ value for the accurate simulation is important. In this study, the y^+ value is recommended to be kept between 1 and 300 [26][30].

The purpose of this paper is to describe the process of developing a two-phase regular wave simulation in OpenFOAM® and obtain a suitable configuration of time step and cell size, which are relevant to the designed wavelength and wave period [31] that strikes a balance between model accuracy and computational time. The model provides a foundation to continue studying the wave attenuation capacity of eco-retrofitted structures, such as the validation for the irregular wave attack and the visualization of the complex wave overtopping process). Thus, a 2D numerical wave tank based on previous experimental work (as documented in [13]) is developed in OpenFOAM®, and the interFoam solver is adopted to generate the desired wave conditions. In this paper, all simulations are run in OpenFOAM v-2212, which was released in 2022 [32].

This paper is organized as follows: The experimental work on which the CFD model is based is described in Section 2, including layout and wave conditions, followed by the CFD governing equations of the transport equation, wave generation, active wave absorption, and $k-\omega$ SST turbulence model in Section 3.1. Then, the detailed process and applied boundary conditions for establishing the 2D numerical wave tank are illustrated in Section 3.2. The study configurations of mesh densities, time steps, and simulation results are described in Section 4. Finally, the main procedures, limitations and recommendations are provided in the conclusion.

2. Experimental dataset

The CFD models developed in this paper were designed to replicate previous experimental data collected in [13]. The scaled model experiments were conducted in the two-dimensional wave channel at the Water Research Facility at the University of Warwick, UK. The flume is 22 m long, 0.6 m wide and 1 m high. A piston-type wavemaker with an active wave absorption system was settled on the left of the wave channel. A plain vertical seawall 0.25 m high, 0.6 m wide and 0.05 m thick was installed 6.25m from the wave paddle on an impermeable foreshore with a uniform slope of 1:2 (Figure 1). Two sets of wave gauges (three in each group) were positioned in the channel to measure the wave height and wave period of the incident waves in relatively deep and shallow water conditions (near structure). An overtopping container suspended from a load cell was installed behind the seawall to collect the overtopping volumes.

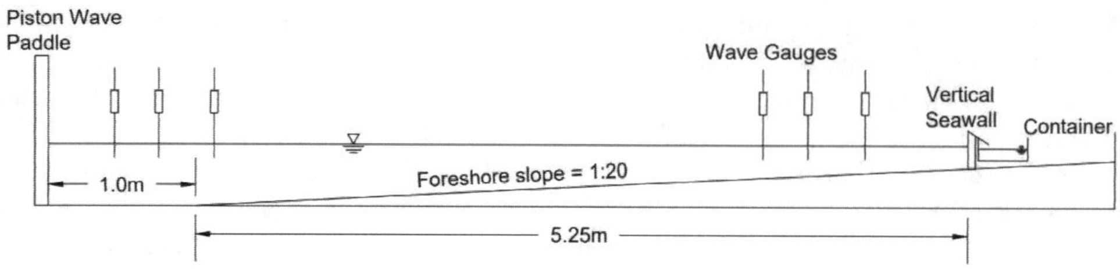

Figure 1: Experimental setup [13]

Each set of experiments comprised approximately 1000 random waves generated with a standard JONSWAP spectrum gamma equaling 3.3. In total, 12 wave conditions were modelled by simulating both impulsive and non-impulsive wave conditions at vertical walls with eco-retrofit configurations. Nevertheless, this study reports only the numerical simulations corresponding to regular waves as a stepping stone for full-scale numerical simulations on eco-retrofitting using irregular sea states. Thus, 0.09m toe water depth, 0.078m wave height and 1.6s wave period are adopted as wave conditions for all the simulation cases.

3. CFD model

This section illustrates the governing equations adopted by OpenFOAM® and the establishment procedures, including boundary conditions.

3.1. Governing equations

An Eulerian mesh-based numerical modelling approach using InterFOAM in conjunction with a k-ω SST turbulence model in OpenFOAM® is employed for the regular wave simulation. Reynolds Averaged Navier–Stokes (RANS) equations are solved by the 'InterFoam' solver for incompressible multiphase (air and water) simulations. The RANS equations include mass continuity(1) and momentum conservation equation (2), which link pressure and velocity, as follows [25].

$$\nabla \cdot \mathbf{U} = \mathbf{0} \tag{1}$$

$$\frac{\partial \rho \mathbf{U}}{\partial t} + \nabla \cdot (\rho \mathbf{U}\mathbf{U}) - \nabla \cdot (\mu_{eff} \nabla \mathbf{U}) = -\nabla p^* - \mathbf{g} \cdot X \nabla \rho + \nabla \mathbf{U} \cdot \nabla \mu_{eff} + \sigma \kappa \nabla \alpha \tag{2}$$

in which, X is the position vector, p^* is the pseudo-dynamic pressure calculated by $p^* = p - \rho \mathbf{g} \cdot \mathbf{X}$, and all bold letters represent vector fields. $\sigma\kappa\nabla\alpha$ is the effect of surface tension, in which σ is the surface tension coefficient, κ is the curvature of the interface calculated by $\kappa = \nabla \cdot \frac{\nabla \alpha}{|\nabla \alpha|}$, and μ_{eff} is the coefficient of dynamic viscosity, which is the sum of molecular dynamic viscosity and turbulent effects: $\mu_{eff} = \mu + \rho v_{turb}$. ρ is the density of each cell calculated by $\rho = \alpha \rho_{\text{water}} + (1-\alpha)\rho_{\text{air}}$, in which, α is the indicator phase function defined as the quantity of water per unit volume at each cell, thus, a cell is full of water if $\alpha = 1$ while the cell is full of air if $\alpha = 0$. Other intermediate values α point to the water-air interface. The properties of the fluid at each cell could then be calculated by the Volume of Fluid (VOF) function.

A modified advection equation is adopted to model the free surface and track the fluid movement [25], as shown in Equation (3):

$$\frac{\partial \alpha}{\partial t} + \nabla \cdot \mathbf{U}\alpha + \nabla \cdot \mathbf{U}_c \alpha(1-\alpha) = 0 \tag{3}$$

in which $\nabla \cdot \mathbf{U}_c \alpha(1-\alpha)$ represents the artificial compression term.

A 2D Active Wave Absorption (AWA) is selected based on the shallow water theory to minimize the influence of reflected waves at the patch boundaries:

$$U_c = -\sqrt{\frac{g}{h}}\eta_R \tag{4}$$

in which U_c is the correction velocity that will be generated as a vector perpendicular to the boundary and point into the domain to cancel out the reflected waves, η_R represents the reflected wave height, which is equal to the difference between the measured wave height at the wavemaker (η_M) and the target elevation (η_T): $\eta_R = \eta_M - \eta_T$.

The $k-\omega$ SST turbulence model is used in this study as it combines the $k-\varepsilon$ model (in the outer region of the boundary layer) and the $k-\omega$ model (in the inner boundary layer) to achieve superior performance [33],[35]. The turbulence specific dissipation rate (ω) and the turbulence kinetic energy (k) are generated by equation (5) and equation (6) respectively [36]-[37]:

$$\frac{\partial}{\partial t}(\rho\omega) = \nabla\cdot(\rho D_\omega \nabla\omega) + \frac{\rho\gamma G}{\nu} - \frac{2}{3}\rho\gamma\omega(\nabla\cdot\mathbf{u}) - \rho\beta\omega^2 - \rho(F_1-1)CD_{\mathbf{k}\omega} + S_\omega \tag{5}$$

$$\frac{\partial}{\partial t}(\rho k) = \nabla\cdot(\rho D_k \nabla k) + \rho G - \frac{2}{3}\rho k(\nabla\cdot\mathbf{u}) - \rho\beta^*\omega k + S_k \tag{6}$$

in which ν_t represents the turbulence viscosity and could be calculated by Equation (7), S_ω and S_k are the user-defined source terms, F_1 is the blending function which is equal to one inside the boundary layer and equal to zero away from the surface corresponding to the $k-\omega$ model and $k-\epsilon$ model respectively.

$$\nu_t = \frac{a_1 k}{\max(a_1\omega, SF_2)} \tag{7}$$

in which, S represents the mean rate of strain of the flow, F_2 is the second blending function.

To deal with two-phase flow simulation, the default solver interFoam and the PIMPLE algorithm (a mixture of PISO and SIMPLE) are adopted [25]. The Courant number (C), as shown in Equation (8), should be at most 1 to ensure that the information will only transfer one cell within each time step; otherwise, the accuracy will be sacrificed. The time step could then be dictated by the Courant number

$$C = \frac{u\times\Delta t}{\Delta x} \tag{8}$$

in which, u is the velocity; Δt represents the maximum allowable time step calculated by the limiting Courant number of 0.5; Δx, in this 2D simulation, denotes the smallest cell size in the velocity direction.

3.2. Modelling procedures

The establishment of a numerical wave tank follows the small-scale wave flume experiments as introduced in Section 2. The computational domain for this 2D test is shown in Figure 2. The thickness of the domain is set to a minimum, i.e., one-cell thick, compared to the other dimensions (0.05m in this case). The sidewalls are set as the empty boundary condition to avoid the fluid movement along the third dimension and the sidewall influence on wave propagation. To create this computational domain, a $1.1m \times 8m \times 1m$ block is generated using the blockMesh utility. Then, the snappyHexMesh tool is applied to insert the geometries of the seawall, container and foreshore slope (.stl file) and then achieve a grid refinement around the structures, resulting in a 3D mesh block. To maintain the normal operation of snappyHexMesh, the boundary conditions of sidewalls are initially set as a “wall” condition. Finally, the extrudeMesh utility is adopted to select the side mesh and extrude it to be one-cell thick and the boundary conditions of sidewalls are converted to “empty” by using changeDictionary utility.

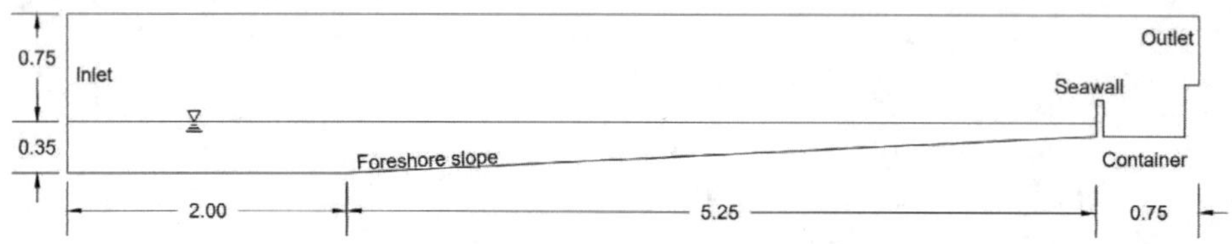

Figure 2: Computational domain for the numerical modelling (all dimensions are in meters)

To increase the performance of snappyHexMesh, there should be at least one intersection of the cell edge with the geometry surface. Thus, considering computational time while ensuring the modelling performance, the element size around the vertical seawall should be set at least 0.05m element size (seawall width). Meanwhile, the mesh densities along the water surface should be refined into the same degree to avoid any errors caused by the inconsistence of grid size and have a better simulation on wave propagation. Thus, the mesh in the vicinity of the geometries is refined three times, resulting in one eighth of the background grid. Furthermore, the simulations are carried out considering the incompressible air and water. The Stokes second-order waves are applied in the numerical wave tank to generate the required waves according to the selected wave conditions.

On the inlet patch, the parameter values of U, k and ω are calculated from the selected wave conditions. The turbulence intensity is calculated by

$$I = \frac{u'}{U} \tag{9}$$

in which, u' is the root-mean-square of the turbulent velocity fluctuations while U represents the mean velocity. The active wave absorption condition is adopted in the waveProperties setting at the inlet and outlet patch to avoid any reflected waves. The no-slip walls and adaptive wall functions are applied for the k and ω configurations in the wall regions. All the parameters are set as empty in the other empty regions. Table 1 summarizes the applied boundary conditions:

Table 1: Summary of the applied boundary conditions

Boundary	Parameters	Type
Inlet patch	alpha.water	waveAlpha
	p_rgh	fixedFluxPressure
	U	waveVelocity
	k	Calculated by $k = \frac{3}{2}(UI)^2$
	nut	Calculated
	omega	Calculated by $\omega = 0.09^{-\frac{1}{4}}\frac{\sqrt{k}}{l}$ where l is turbulence length-scale, normally using 0.07 in this case
Outlet and atmosphere patch	alpha.water	inletOutlet
	p_rgh	totalPressure
	nut	Calculated
	U	pressureInletOutletVelocity
	k, omega	inletOutlet
Wall	alpha.water	Zero surface-normal gradient
	p_rgh	fixedFluxPressure
	U	No-slip
	k, omega, nut	Adaptive wall function
Empty	alpha.water, p_rgh, U, k, omega, nut	Empty

4. Investigation of grid resolution and time resolution

Regular waves are generated with various configurations of mesh densities and time resolutions. Then, the simulated water elevation variations at the 2m (deep water condition) and 6.55m (near-structure and shallow water condition) from the wave paddle are compared with each other. Model settings and corresponding results are presented in the following sections.

4.1. Grid resolution

Model settings. The initial coarse mesh is established based on the rule of thumb that $\Delta x = L/320$; where L is the wavelength and Δx is the smallest cell size. With the limits of guideline Courant numbers, a time step of 0.001s is initialized first. Then, three meshes with different levels of refinement are generated using $L/460$, $L/600$, and $L/800$ respectively, as shown in Table 2:

Table 2: Different configurations of grid resolutions

Test series	Grid description	Min element size (m)	Max element size (m)	Number of elements
Case 1	L/320	0.0125	0.1	31146
Case 2	L/460	0.009	0.07	59999
Case 3	L/600	0.0067	0.05	105517
Case 4	L/800	0.005	0.04	175034

It is worth mentioning that the average y^+ value at the wall varies from 125 (Case 1), 75 (Case 2), 50 (Case 3) and down to 25 (Case 4), which are all within the optimal range of y^+ values (1 to 300) when employing the wall functions in simulations.

Results. As shown in Figure 3, firstly, all the cases have a wave period of 1.6s, which is 0.05s smaller than the input value. Secondly, the average wave heights for Case 1, Case 2, Case 3 and Case 4 are 0.079463m, 0.080213m, 0.078306m and 0.083121m, respectively, leading to errors of 1.875%, 2.837%, 0.3923% and 6.565% when compared to the input wave height of 0.078m. Besides, the arrival of waves in Case-4 has a considerable delay compared with the other three groups, which may be caused by unexpected parameter errors and should be weeded out.

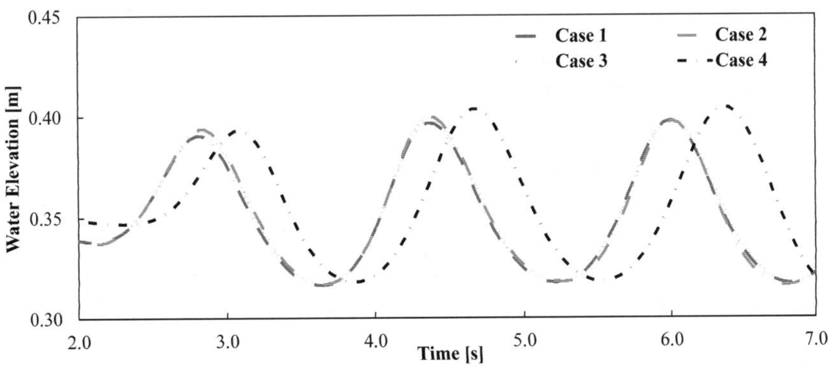

Figure 3: A grid resolution study and its comparison of water elevation variation at 2m from the wave paddle

When comparing the data at the near-structure position, as shown in Figure 4, it is found that the differences between research cases are higher than in Figure 3, which reflects the importance of mesh density in dealing with the incident waves and reflected waves. The simulated wave periods for Case 1, Case 2 and Case 3 are 1.625s, 1.625s and 1.6125s, respectively, leading to errors of 1.56%, 1.56% and

0.78% compared to the input wave period of 1.6s. Based on these, the mesh density of L/600 (Grid case–3) is selected for the time resolution study.

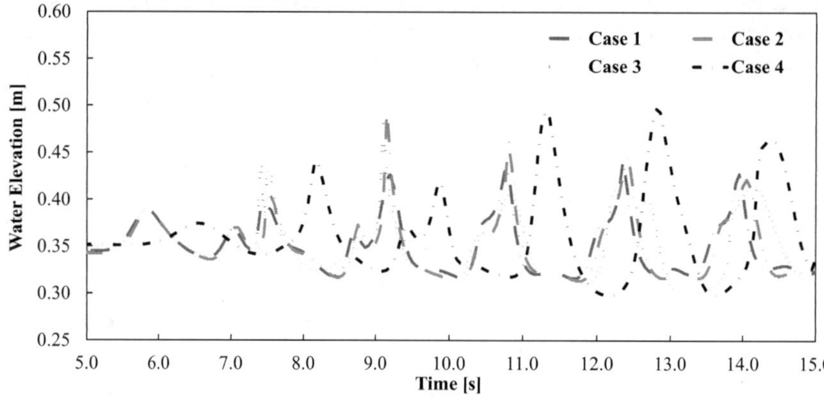

Figure 4: A grid resolution study and its comparison of water elevation variation at 6.55m from the wave paddle (near-structure)

4.2. Time resolution

Model settings. From the previous introduction, the maximum time step is determined by the Courant number (0.5), wave condition, and selected cell size (here, L/600). Then, the other three sets of time steps are conducted by the wave period (T), as shown in the Table 3:

Table 3: Different configurations of time resolutions

Maximum time step (s)	Test series	Description	Time step (s)
0.00133	Case 5	$\Delta t = T/1600$	0.001
	Case 6	$\Delta t = T/3000$	0.00053
	Case 7	$\Delta t = T/6000$	0.00027
	Case 8	$\Delta t = T/12000$	0.00013

Results. As shown in Figure 5, the change in time steps almost has no influence on wave generation in deep water conditions. The simulated wave height and wave period are 0.07833m and 1.575s, respectively.

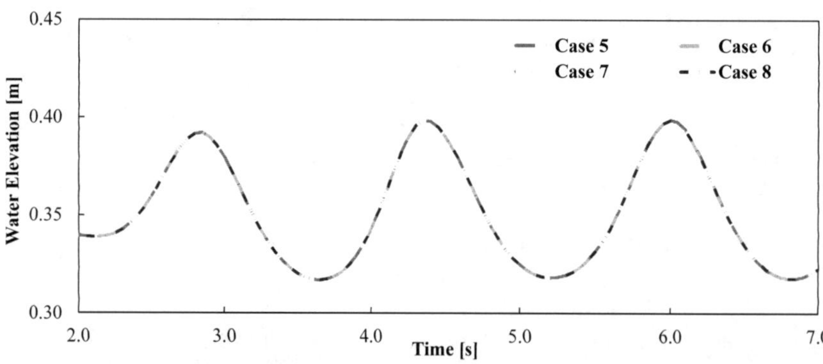

Figure 5: A time resolution study and its comparison of water elevation variation at 2m from the wave paddle

However, as shown in Figure 6, different time steps still make a big difference when dealing with the breaking waves near the structure. It is found that the water elevation of Case 8 at 9.1s is 0.06m lower than the other three cases, while the water elevation of Case 6 at 12.5s is nearly 0.1m higher than the different cases. This may be due to the unexpected errors in parameter setting, thus, Case 8 and Case 6 should be excluded. There are no significant differences when comparing the simulated results of Case 5 and Case 7. Therefore, the time step of Case 5 (T/1600) is recommended for further study as Case 7 needs more computational cost without extra benefits.

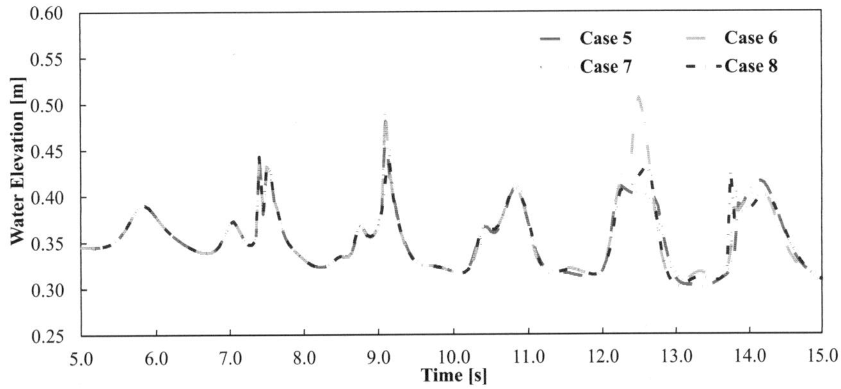

Figure 6: A time resolution study and its comparison of water elevation variation at 6.55m from the wave paddle (near-structure)

5. Conclusion

Facing the increasing requirements of developing 'greener' sea defences with better climate resilience, eco-retrofitted structures that combine traditional sea defences and biological characteristics to provide dual benefits of wave attenuation and ecological remediation have gained more and more attention as an innovative intervention for existing structures. However, the wave attenuation mechanism of such interventions is still underdeveloped.

As the first step to model the wave attenuation performance of eco-retrofitted structures, this paper presents an entire process of generating two-phase regular wave simulation using the interFoam solver in OpenFOAM®, an open-source computational fluid dynamics framework. A 2D numerical wave tank is established according to the scaled experimental settings to study the influence of mesh density and time step. The performances of water elevation variations at deep and shallow water conditions (near-structure) are discussed and compared with the input conditions. The following key points can be drawn from this work:

1. InterFoam solver within OpenFOAM® has a reliable ability to generate the desired wave conditions.
2. In deep water conditions, the changes in time step almost have no significant influence on the generated wave parameters.
3. In the shallow water condition near the structure, the influence of mesh density is much more significant than in the deep water condition. This is likely due to the complicated dynamics of the wave-breaking process in the wave reflection area. Thus, applying a finer mesh in shallow water conditions (near-structure) is recommended to study the wave-structure interaction, which leads to understanding the wave dissipation mechanisms of eco-retrofitted structures.
4. Mesh density has a higher influence on the simulation accuracy than the time step. Generally, there won't be significant differences once the time step has satisfied the maximum requirements of the Courant number because it has avoided the information propagating through more than one grid cell at each time step.
5. Several unexpected simulation results have been observed in this work (e.g., the unexpected wave delay in Case 4), which might be due to the unsuitable parameter settings in the model.

6. Based on the modelling performance and computational time, wave profiles at $\Delta x = L/600$ generated with $\Delta t = T/1600$ (average y^+ value around 50) are suggested for future studies to reproduce and validate the irregular waves generated in the experiments.

Acknowledgement
Xihang Xu gratefully acknowledges the China Scholarship Council and UCD School of Civil Engineering for the financial support of his PhD study. We would also like to sincerely thank the organizing committee of the 18th OpenFOAM Workshop for their support and advise in developing this study.

Statement
Numerical models as applied in this study are currently under further development, which requires further validation. As such, the case/source or code/data have not been made open access.

References
[1] Singhvi, A., Luijendijk, A.P., van Oudenhoven, A.P.E., 2022. The grey – green spectrum: A review of coastal protection interventions. Journal of Environmental Management 311, 114824.
[2] Salauddin, M., Pearson, J., 2018. A laboratory study on wave overtopping at vertical seawalls with a shingle foreshore. Coastal Engineering Proceedings 56–56.
[3] Van den Hoven, K., Kroeze, C., van Loon-Steensma, J.M., 2022. Characteristics of realigned dikes in coastal Europe: Overview and opportunities for nature-based flood protection. Ocean & Coastal Management 222, 106116.
[4] Kiesel, J., MacPherson, L.R., Schuerch, M., Vafeidis, A.T., 2022. Can Managed Realignment Buffer Extreme Surges? The Relationship Between Marsh Width, Vegetation Cover and Surge Attenuation. Estuaries and Coasts 45, 345–362.
[5] Chang, C.-W., Mori, N., 2021. Green infrastructure for the reduction of coastal disasters: a review of the protective role of coastal forests against tsunami, storm surge, and wind waves. Coastal Engineering Journal 63, 370–385.
[6] Scheres, B., Schüttrumpf, H., 2019. Nature-Based Solutions in Coastal Research – A New Challenge for Coastal Engineers?, in: Trung Viet, N., Xiping, D., Thanh Tung, T. (Eds.), APAC 2019. Springer, Singapore, pp. 1383–1389.
[7] Hall, A.E., Herbert, R.J.H., Britton, J.R., Boyd, I.M., George, N.C., 2019. Shelving the Coast With Vertipools: Retrofitting Artificial Rock Pools on Coastal Structures as Mitigation for Coastal Squeeze. Frontiers in Marine Science 6.
[8] Morris, R.L., Konlechner, T.M., Ghisalberti, M., Swearer, S.E., 2018. From grey to green: Efficacy of eco-engineering solutions for nature-based coastal defence. Global Change Biology 24, 1827–1842.
[9] Firth, L.B., Schofield, M., White, F.J., Skov, M.W., Hawkins, S.J., 2014. Biodiversity in intertidal rock pools: Informing engineering criteria for artificial habitat enhancement in the built environment. Marine Environmental Research, Special Issue: Managing Biodiversity in a Changing Ocean 102, 122–130.
[10] Evans, A.J., Firth, L.B., Hawkins, S.J., Morris, E.S., Goudge, H., Moore, P.J., Evans, A.J., Firth, L.B., Hawkins, S.J., Morris, E.S., Goudge, H., Moore, P.J., 2015. Drill-cored rock

pools: an effective method of ecological enhancement on artificial structures. Mar. Freshwater Res. 67, 123–130.

[11] Strain, E.M.A., Olabarria, C., Mayer-Pinto, M., Cumbo, V., Morris, R.L., Bugnot, A.B., Dafforn, K.A., Heery, E., Firth, L.B., Brooks, P.R., Bishop, M.J., 2018. Eco-engineering urban infrastructure for marine and coastal biodiversity: Which interventions have the greatest ecological benefit? Journal of Applied Ecology 55, 426–441.

[12] Sawyer, A.C., Toft, J.D., Cordell, J.R., 2020. Seawall as salmon habitat: Eco-engineering improves the distribution and foraging of juvenile Pacific salmon. Ecological Engineering 151, 105856.

[13] Salauddin, M., O'Sullivan, J.J., Abolfathi, S., Pearson, J.M., 2021. Eco-Engineering of Seawalls—An Opportunity for Enhanced Climate Resilience From Increased Topographic Complexity. Frontiers in Marine Science 8.

[14] O'Sullivan, J., Salauddin, M., Abolfathi, S., Pearson, J., 2020. Effectiveness of eco-retrofits in reducing wave overtopping on seawalls. Coastal Engineering Proceedings 13–13.

[15] Salauddin, M., O'Sullivan, J.J., Abolfathi, S., Peng, Z., Dong, S. and Pearson, J.M., 2022. New insights in the probability distributions of wave-by-wave overtopping volumes at vertical breakwaters. Scientific Reports, 12(1), p.16228.

[16] Habib, M.A., O'Sullivan, J.J., Abolfathi, S. and Salauddin, M., 2023. Enhanced wave overtopping simulation at vertical breakwaters using machine learning algorithms. Plos one, 18(8), p.e0289318.

[17] Yeganeh-Bakhtiary, A., Houshangi, H., Hajivalie, F., Abolfathi, S., 2018. A numerical study on hydrodynamics of standing waves in front of caiss on breakwaters with WCSPH model. Coastal Engineering Journal 59, 1750005–1.

[18] Qu, K., Huang, J.X., Guo, L., Li, X.H., 2022. Numerical Study on Hydrodynamics of Submerged Permeable Breakwater under Impacts of Focused Wave Groups Using a Nonhydrostatic Wave Model. Journal of Marine Science and Engineering 10, 1618.

[19] Altomare, C., Crespo, A.J.C., Domínguez, J.M., Gómez-Gesteira, M., Suzuki, T., Verwaest, T., 2015. Applicability of Smoothed Particle Hydrodynamics for estimation of sea wave impact on coastal structures. Coastal Engineering 96, 1–12.

[20] Chen, W., Warmink, J.J., van Gent, M.R.A., Hulscher, S.J.M.H., 2021. Numerical modelling of wave overtopping at dikes using OpenFOAM®. Coastal Engineering 166, 103890.

[21] Irías Mata, M., van Gent, M.R.A., 2023. Numerical modelling of wave overtopping discharges at rubble mound breakwaters using OpenFOAM®. Coastal Engineering 181, 104274.

[22] Wang, B., Li, Y., Wu, F., Gao, S., Yan, J., 2022. Numerical Investigation of Wave Run-Up and Load on Fixed Truncated Cylinder Subjected to Regular Waves Using OpenFOAM. Water 14, 2830.

[23] Mohseni, M., Esperanca, P.T., Sphaier, S.H., 2018. Numerical study of wave run-up on a fixed and vertical surface-piercing cylinder subjected to regular, non-breaking waves using OpenFOAM. Applied Ocean Research 79, 228–252.

[24] Ravindar, R., Sriram, V., Salauddin, M., 2022. Numerical modelling of breaking wave impact loads on a vertical seawall retrofitted with different geometrical configurations of recurve parapets. Journal of Water and Climate Change 13, 3644–3674.

[25] Higuera, P., Lara, J.L., Losada, I.J., 2013. Realistic wave generation and active wave absorption for Navier–Stokes models: Application to OpenFOAM®. Coastal Engineering 71, 102–118.

[26] Devolder, B., Rauwoens, P., Troch, P., 2017. Application of a buoyancy-modified k-ω SST turbulence model to simulate wave run-up around a monopile subjected to regular waves using OpenFOAM®. Coastal Engineering 125, 81–94.

[27] Adanta, D., Fattah, I.M.R., Muhammad, N.M., 2020. Comparison of standard k-epsilon and sst k-omega turbulence model for breastshot waterwheel simulation. Journal of Mechanical Science and Engineering 7, 039–044.

[28] Brown, S.A., Magar, V., Greaves, D.M., Conley, D.C., 2014. An evaluation of RANS turbulence closure models for spilling breakers. Coastal Engineering Proceedings 5–5.

[29] Lodh, B.K., Das, A.K., Singh, N., 2017. Investigation of turbulence for wind flow over a surface mounted cube using Wall Y^+ approach. Indian J. Sci. Technol 10, 1–11.

[30] Wang, B., Li, Y., Wu, F., Gao, S., Yan, J., 2022. Numerical Investigation of Wave Run-Up and Load on Fixed Truncated Cylinder Subjected to Regular Waves Using OpenFOAM. Water 14, 2830.

[31] Lara, J.L., Higuera, P., Guanche, R., Losada, I.J., 2013. Wave Interaction With Piled Structures: Application With IH-FOAM. Presented at the ASME 2013 32nd International Conference on Ocean, Offshore and Arctic Engineering, American Society of Mechanical Engineers Digital Collection.

[32] OpenFOAM GitLab, 2023. URL:https://develop.openfoam.com/Development/openfoam/-/wikis/precompiled/debian.

[33] Devolder, B., Rauwoens, P., Troch, P., 2017. Application of a buoyancy-modified k-ω SST turbulence model to simulate wave run-up around a monopile subjected to regular waves using OpenFOAM®. Coastal Engineering 125, 81–94.

[34] Adanta, D., Fattah, I.M.R., Muhammad, N.M., 2020. Comparison of standard k-epsilon and sst k-omega turbulence model for breastshot waterwheel simulation. Journal of Mechanical Science and Engineering 7, 039–044.

[35] Rahman, M.M., Karim, M.M., Alim, M.A., 2007. Numerical investigation of unsteady flow past a circular cylinder using 2-D finite volume method. Journal of Naval Architecture and Marine Engineering 4, 27–42.

[36] Menter, F.R., Kuntz, M., Langtry, R., 2003. Ten years of industrial experience with the SST turbulence model. Turbulence, heat and mass transfer 4, 625–632.

[37] OpenFOAM: User Guide: k-omega Shear Stress Transport (SST), URL: https://www.openfoam.com/documentation/guides/latest/doc/guide-turbulence-ras-k-omega-sst.html.

IOP Conf. Series: Materials Science and Engineering **1312** (2024) 012007
doi:10.1088/1757-899X/1312/1/012007

Backward facing step: from fluid flow to conjugate heat transfer with the coupling library preCICE

C G Caccia[1], M Corti[2], A Della Torre[2], and P Masarati[1]

[1]Politecnico di Milano, Milan, Italy, Department of Aerospace Science and Technology
[2]Politecnico di Milano, Milan, Italy, Department of Energy

E-mail: claudiogiovanni.caccia@polimi.it

Abstract. The Backward Facing Step geometry is a widely used benchmark problem in Computational Fluid Dynamics literature that is exploitable to validate models, solution methods, and software implementations. Despite a simple geometry, it shows phenomena like separation, reattachment, and re-circulation zones, under different flow conditions (i.e. different Reynolds number or turbulence parameters) it gives different measurable results, suitable for benchmarking activities [1]. Also regarding heat transfer analysis, the backward facing step can be used to investigate a wide variety of operating conditions (both for simple heat transfer cases and coupling heat transfer between the fluid region and a neighboring solid region giving rise to a more complex conjugate heat transfer model) [2]. This work uses the backward facing step as a test case to validate a numerical model built with the open-source Software OpenFOAM 10. The fluid and solid subdomains are connected through the open-source coupling library preCICE [3]. The results, taken from simulations carried out by the authors, show good agreement with the data available in the literature.

1. Introduction

The Backward Facing Step (BFS) geometry is a widely used benchmark problem in Computational Fluid Dynamics (CFD) literature to validate models, solution methods, and software implementations, mainly for the following reasons: despite a simple geometry, it shows separation, reattachment, and recirculation zones [4]; different flow conditions (i.e. different Reynolds numbers, Re) give different measurable results, suitable for benchmarking; heat flux from the lower wall can be used to investigate heat transfer characteristics; coupling heat transfer between the fluid region and a neighboring solid region gives rise to a conjugate heat transfer model. [5]

The BFS flow has the advantage of being described by a simple geometry and showing enough complexity to be considered as a test case. It is thus worth studying from many different points of view, based on the specific physics involved. Figure 1 shows a typical configuration of a backward facing step, introducing the main geometric parameters.

2. Objectives

The objectives of the present activity consist of generating a case within the OpenFOAM 10 framework for analyzing various aspects of the BFS model. Regarding fluid flow, we aim to evaluate the fluid model performance concerning pressure and velocity fields. Regarding heat transfer, we introduce in the model the energy equation, this stage is useful as an intermediary

IOP Conf. Series: Materials Science and Engineering **1312** (2024) 012007 doi:10.1088/1757-899X/1312/1/012007

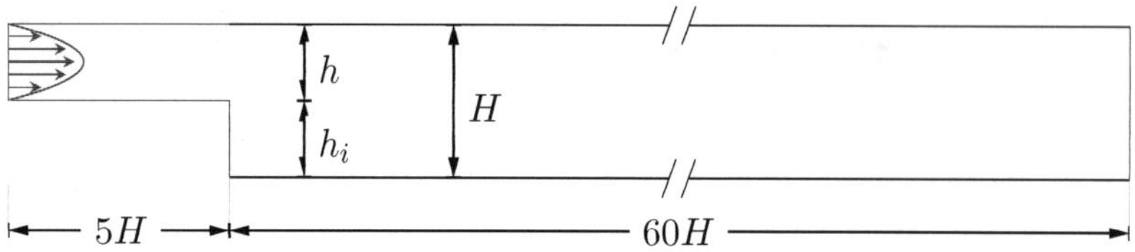

Figure 1. BFS geometry.

model, focusing on key heat transfer parameters. Regarding Conjugate Heat Transfer (CHT), we propose an integrated model consisting of a fluid region (validated in prior steps) and a solid region, where temperature fields and heat flux remain continuous at the interface. Throughout each stage, comparisons with existing literature data have been conducted to validate the findings.

3. Fluid flow simulations

Many studies have been conducted on the BFS. Considering an incompressible flow, the Reynolds number (based on the maximum channel height) is the driving parameter, and simulations range from $Re = 50$ to and over 80.000 [1].

In this study, we decided to focus, for the fluid flow analysis, on the case with:

$$Re_H = Re_{2h} = 800 \tag{1}$$

that is the case in which the step height is equal to the inlet height and the channel height ($h = h_i$ in Figure 1).

The CFD simulation is defined by a set of initial conditions encompassing geometric parameters and fluid properties, consolidated within a single configurable file. This file dictates geometric properties, mesh discretization, and essential fluid attributes such as velocity and viscosity calibrated to achieve a Reynolds number of $Re_H = 800$ (Equation 1).

A uniform mesh, generated using `blockMesh`, is employed, while boundary conditions are defined as follows: the inlet features a parabolic velocity profile, the outlet is set to uniform pressure and walls use a no-slip velocity condition.

Steady and incompressible analysis is conducted applying the `simpleFoam` solver, with velocity and pressure residuals set at 10^{-6} and 10^{-4}, respectively, to define the convergence of the simulation.

The results provided in various studies, such as [6], [7] and [5], allow assessing the fluid-dynamic quality of the model using the following parameters for comparison:

(i) Reattachment regions:
- $\frac{X_1}{h}$: first reattachment point on the lower wall
- $\frac{X_2}{h}$: first separation point on the upper wall
- $\frac{X_3}{h}$: first reattachment point on the upper wall

These regions are shown in Figure 3 and 4 for a better understanding.

(ii) Velocity and pressure profiles at section $\frac{X}{h} = 14$ and $\frac{X}{h} = 30$

The previous parameters have been used to perform a grid independence study.

The position of the separation and reattachment points (see Figures 3 and 4) is analyzed as a function of the mesh density, measured as the number of control volumes (CV) along h (volumes are square). The results are shown in Figure 5: all separation and reattachment points are within the benchmark region (see [7]) using a mesh with around 100 control volumes.

Data published in [6] allow to compare, once made non-dimensional, the values of the velocity and pressure profiles computed at different sections, in particular at $x = 14h$ and $x = 30h$. Figure 6 shows that the higher the control volumes, the closer the results are to the benchmark provided in [6]. Again, a number of control volumes around 100 looks to be enough to correctly describe the fluid flow.

4. Coupling with preCICE

The preCICE [3] software library[1], is an open-source tool designed to implement the connection of single-physics solvers, enabling the creation of partitioned multi-physics simulations such as fluid-structure interaction and conjugated heat transfer among others.

The simulations presented here involve the multiphysics coupling library preCICE [3], exploiting the adapter between preCICE and OpenFOAM [8].

Performing conjugate heat transfer (CHT) simulations using preCICE and OpenFOAM, rather than relying solely on `chtMultiRegionFoam`, offers several advantages. First, it provides better control over the coupling process between fluid and solid domains. This greater control facilitates the exploration of various coupling strategies and parameter settings, leading to improved accuracy and reliability in predicting heat transfer phenomena. Moreover, this allowed us to gain experience and insight into CHT parameters, fostering a deeper understanding of the underlying physics and computational techniques involved. Furthermore, the adoption of preCICE and OpenFOAM opens doors to future possibilities, such as coupling with other solid solvers, allowing to define different physical properties or to retrieve thermal stresses and deformations.

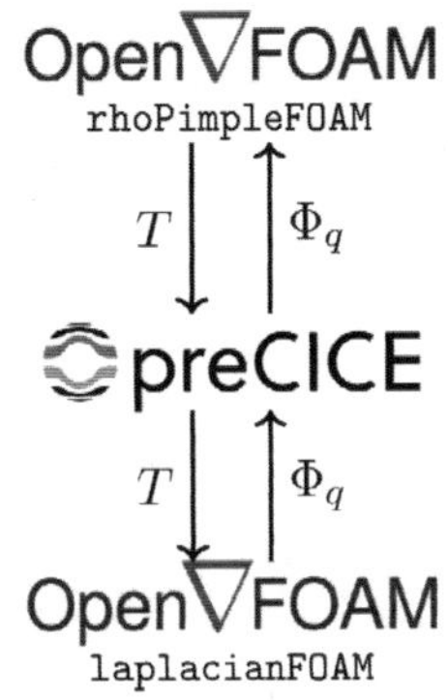

Figure 2. Coupling scheme.

The preCICE coupling library enables communication between different physics solvers (in our case `rhoPimpleFoam` and `laplacianFoam`, as depicted in Figure 2). Its functionality extends to the definition of coupling strategies, with options like *loose* versus *tight* or *staggered* versus *parallel* approaches. Besides, preCICE allows the configuration of convergence methods and criteria, to improve the efficiency of coupled simulations. In case of non-conformal meshes at the interface of the domains, it is able to map data across disparate computational domains.

1 the code is accessible at the following repository: github.com/precice/precice

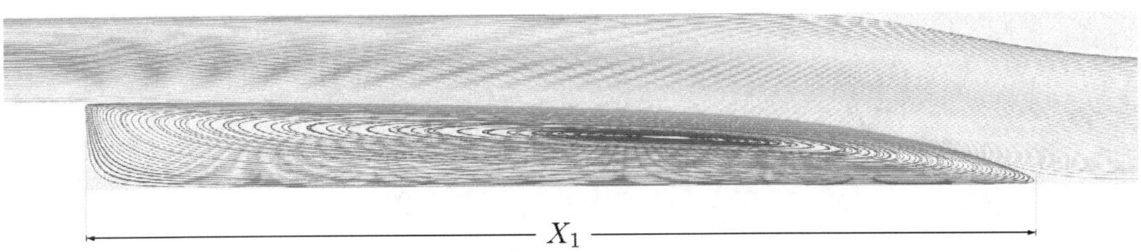

Figure 3. X_1 - first reattachment point.

IOP Conf. Series: Materials Science and Engineering **1312** (2024) 012007 doi:10.1088/1757-899X/1312/1/012007

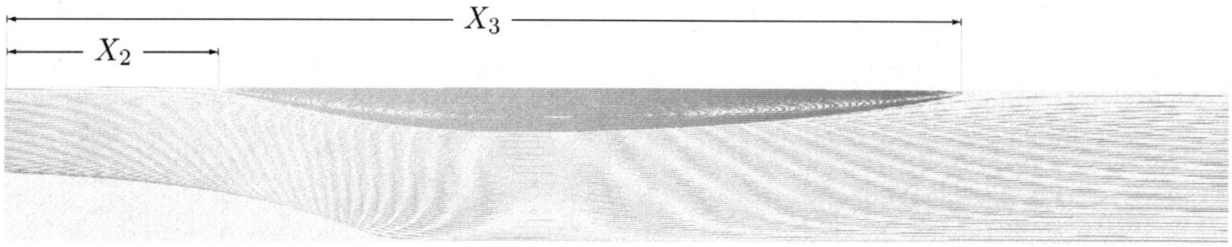

Figure 4. X_2 and X_3 - upper separation and reattachment points.

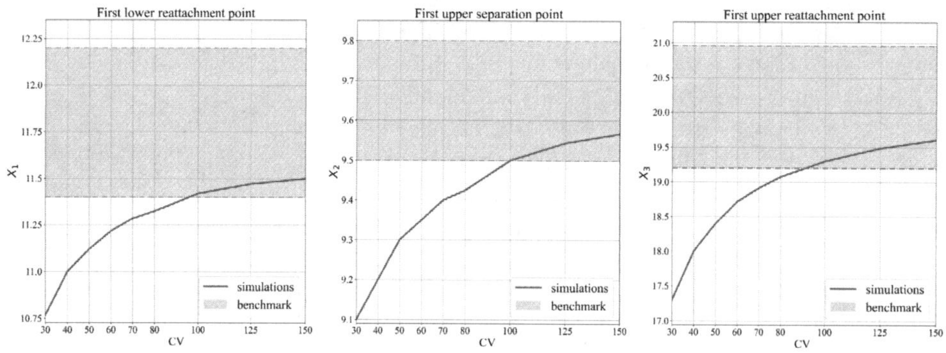

Figure 5. Separation and reattachment points.

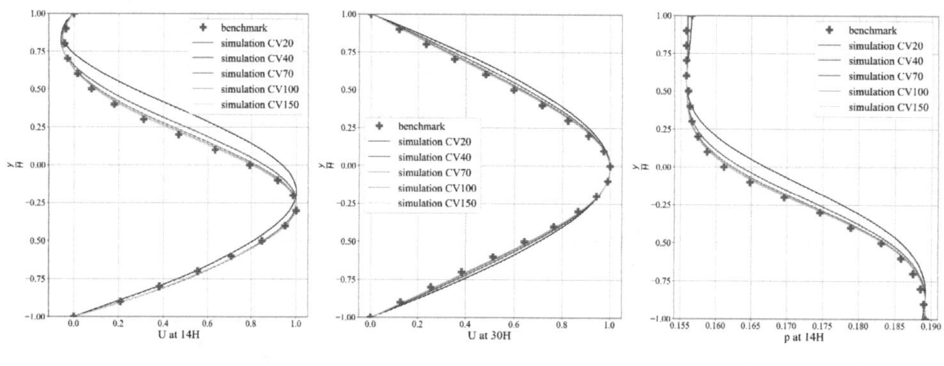

Figure 6. Velocity and pressure profiles.

5. Conjugate Heat Transfer simulations

The Conjugate Heat Transfer analysis stems from the fluid domain analyzed in Section 3 coupled with a solid domain. The two regions share a surface on which they exchange data (see Figure 7). The interface boundary conditions are expressed by the following equations:

$$T_s|_{interf} = T_f|_{interf} \tag{2}$$

$$k_s \frac{\partial T_s}{\partial y}\bigg|_{interf} = -k_f \frac{\partial T_f}{\partial y}\bigg|_{interf} \tag{3}$$

Equation 2 states that the temperature is continuous at the interface, while Equation 3 states that the heat flux exiting one region is equal to the heat flux entering the other region.

IOP Conf. Series: Materials Science and Engineering **1312** (2024) 012007 doi:10.1088/1757-899X/1312/1/012007

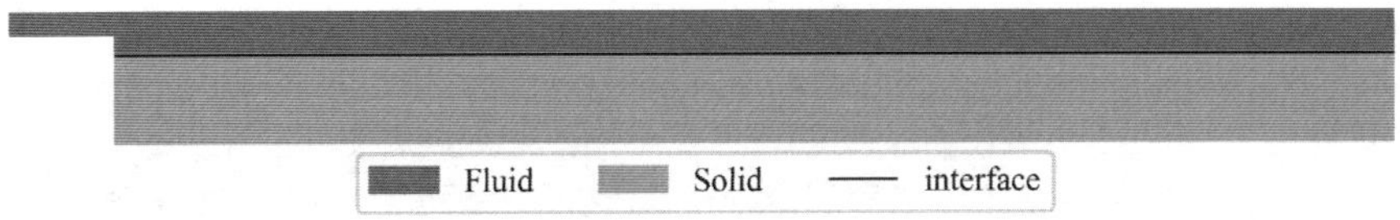

Figure 7. CHT domain.

5.1. Model Parameters

The CFD model has been built considering the following parameters:

Mesh
Same mesh for the fluid domain as previous cases, mesh for the solid domain connected to the lower part of the channel, with height $b = 4h$. Mesh defined using `blockMesh`.

Initial conditions
Fluid flow initialization as previous cases, solid initialized with uniform temperature T_{inlet}, while the lower wall of the solid is initialized with T_{lw}.

Fluid boundary and initial conditions
Same BCs as previous cases, in the domain *steady state* fluid solution and uniform temperature T_{inlet}.

Solid boundary and conditions
Imposed temperature T_{lw} at lower wall, adiabatic side walls, uniform temperature T_{inlet} in the domain.

Interface
Conditions expressed in Equation 2 and 3.

Solvers
`rhoPimpleFoam` and `laplacianFoam` with suitable Δt in order to have a reasonable *Courant number* in the fluid region and *diffusion* number in the solid region.

Physical Properties

- `fluid` region: $Re_H = 800$, $Pr = 0.71$, density $\rho = const$
- `solid` region: thermal conductivity k_S such that $k_{ratio} = \frac{k_S}{k_F} = [1, 10, 100, 1000]$.

5.2. Benchmark data

The data available in the literature (see [4] and [5]) to check the quality of the solution are the temperature distribution in the two regions at defined sections ($\frac{x}{H} = 6, 14, 30$), the temperature profile and the Nusselt number along the interface.
The local Nusselt number is defined as follows:

$$Nu(x) = -\left.\frac{\partial\theta}{\partial n}\right|_{interf} \qquad (4)$$

where θ is the dimensionless temperature, defined as $\theta = \frac{T - T_{inlet}}{T_{lw} - T_{inlet}}$

6. Results

We performed transient simulations on temperature starting from the fully developed fluid flow to assess the temperature evolution in the fluid and in the solid, upon reaching a steady state. We measured the temperature evolution at 4 watchpoints on the interface, as shown in Figure 8. Figure 9 shows the evolution of the temperature at different values of k_{ratio}, which affects the final temperature at watchpoints and the time required to reach a steady state condition.

IOP Conf. Series: Materials Science and Engineering **1312** (2024) 012007 doi:10.1088/1757-899X/1312/1/012007

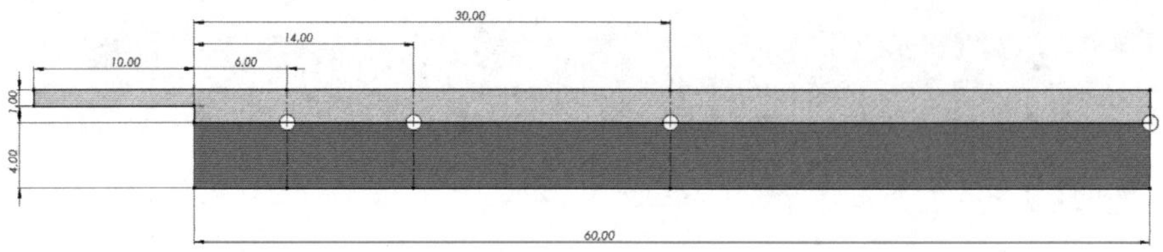

Figure 8. Watch-points.

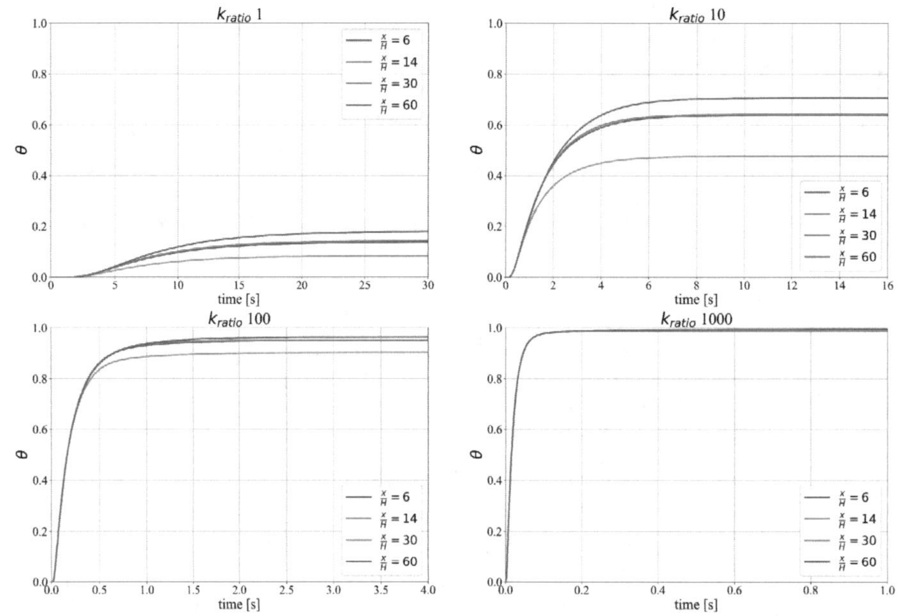

Figure 9. Temperature distribution.

6.1. Temperature profiles on sections

Figures 10 and 11 show the temperature profiles, taken at different positions, for the numerical case compared with the benchmark data. The temperature profiles look identical to the benchmark. The strong coupling imposed in the simulations looks to guarantee the expected solution. It's possible to notice that the continuity of the solution between the solid and the fluid domain is ensured.

6.2. Temperature and Nusselt number profiles at interface

Figures 12 and 13 show the temperature profiles, taken at the interface, for the numerical case solutions compared with the benchmark data. Similarly to the previous results, in each figure the value of the k_{ratio} parameters changes between the proposed values. Only the cases with $k_{ratio} = 1, 10$ are shown.

The graphs regarding the temperature along the interface confirm good agreement with the benchmark, although the lowest temperature position and value is slightly underestimated, in particular for $k_{ratio} = 100$. The Nusselt number (expression of the heat flux at the interface), shown in Figures 14 and 15, is very close to the benchmark, apart from the peak region corresponding to the recirculation zone, where the temperature is slightly underestimated.

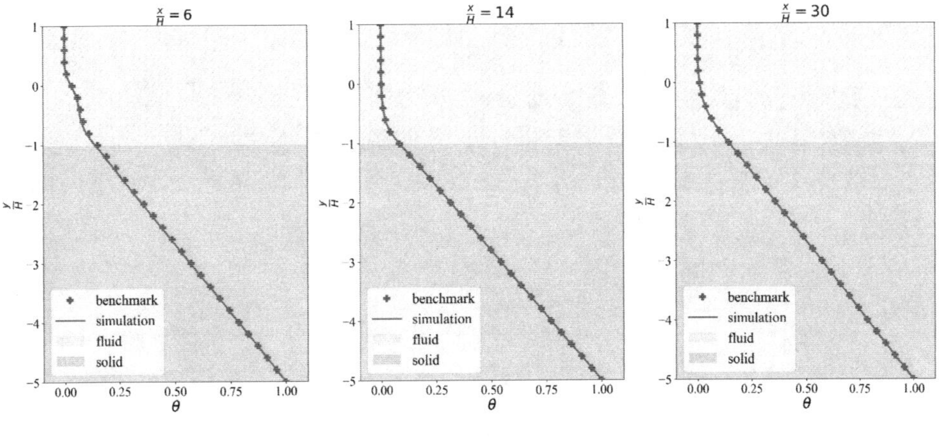

Figure 10. Temperature profile for $k_{ratio} = 1$.

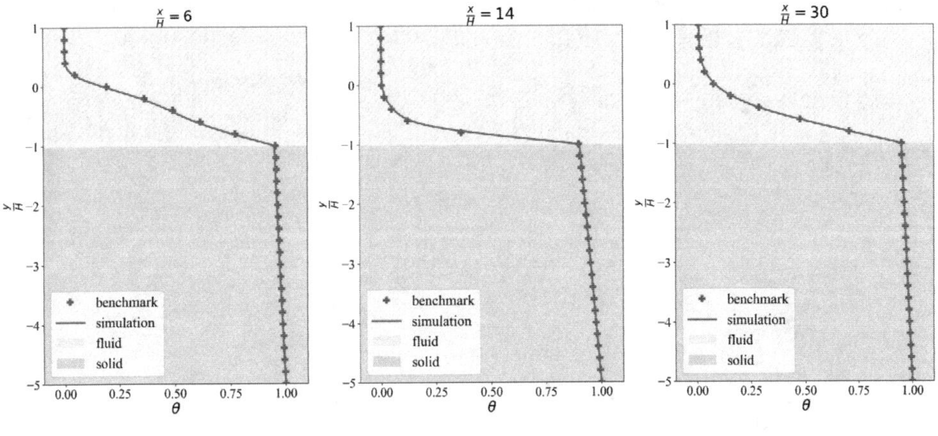

Figure 11. Temperature profile for $k_{ratio} = 100$.

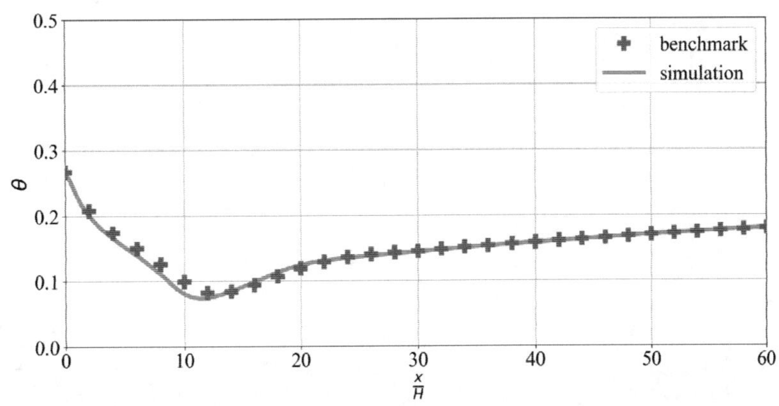

Figure 12. Temperature along interface for $k_{ratio} = 1$.

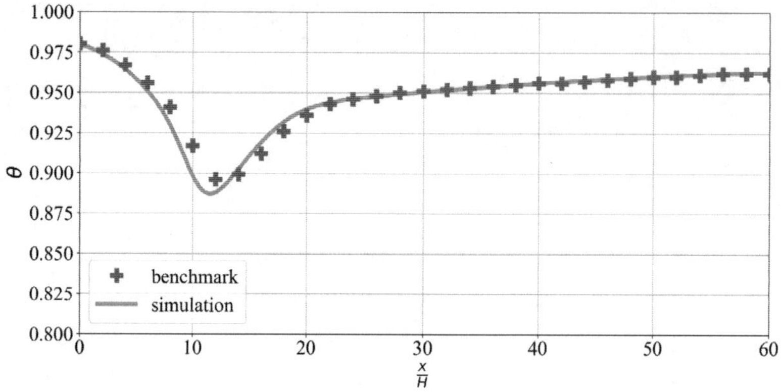

Figure 13. Temperature along interface for $k_{ratio} = 100$.

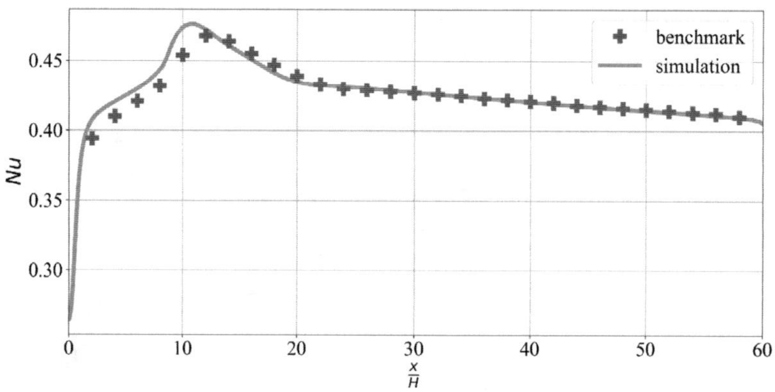

Figure 14. Nusselt number along interface for $k_{ratio} = 1$.

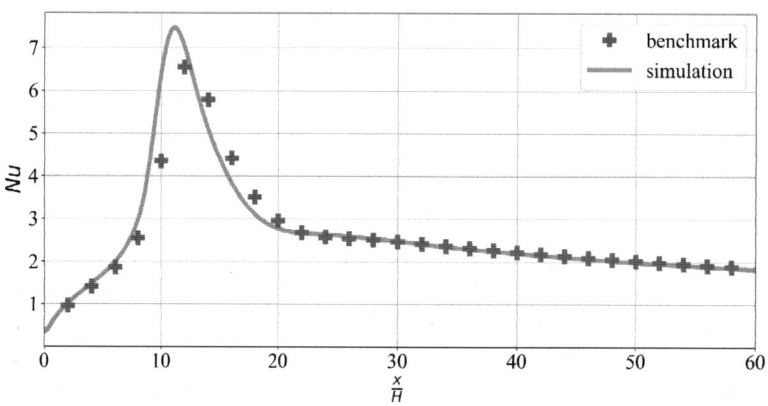

Figure 15. Nusselt number along interface for $k_{ratio} = 100$.

6.3. Coupling parameters

The coupling model concerns the way the solvers are interconnected and how convergence is reached. The configuration for our simulations, defined in preCICE, considers:

- *strong coupling*: that is, each time-step is iterated until convergence is reached,

- *staggered simulation*: `rhoPimpleFoam` is executed first, `laplacianFoam` second,
- *Aitken* under-relaxation [9]: other more complex acceleration proved no better,
- *residual control*: relative measure $1 \cdot 10^{-4}$ on temperature and heat flux.

the parameter k_{ratio} impacted the average number of iterations required to reach convergence, as shown in Table 1:

Table 1. Average coupling iterations.

k_{ratio}	1	10	100	1000
avg. iterations	1.389	2.871	6.910	21.216

7. Conclusions

In this proposed analysis, different aspects concerning fluid flow and heat transfer in a BFS geometry have been validated. For the fluid-dynamics aspects, the focus has been on a chosen Reynolds number ($Re_H = 800$), in which the flow is laminar but close to transition. A relationship between the mesh fineness and the main separation and reattachment points of the flow have been studied. The analysed characteristics have been the reattachment regions and velocity and pressure profiles. All the considered parameters show good agreement with literature data.

For the conjugate heat transfer analysis a model consisting in a solid domain attached to the lower wall of the channel has been considered. In particular, the temperature distribution in the solid and in the fluid regions as a function of the ratio between the fluid and the solid thermal conductivity have been analyzed. All models have been created using different OpenFOAM solvers connected through the coupling library preCICE.

All simulations showed satisfactory agreement between numerical results and the benchmarks data available in literature.

The BFS geometry, despite its simplicity, confirms to be an affordable candidate to validate models ranging from pure CFD to more complex multiphysics models like CHT.

Availability of data and materials

The dataset and the materials used for the study described in this article is available upon request to the authors.

References

[1] Chen L, Asai K, Nonomura T, Xi G and Liu T 2018 *Therm. Sci. Eng. Prog.* **6** 194–216
[2] Kanna P R and Das M K 2006 *Int. J. heat mass transf.* **49** 3929–3941
[3] Chourdakis G, Davis K, Rodenberg B, Schulte M, Simonis F, Uekermann B, Abrams G, Bungartz H, Cheung Yau L, Desai I, Eder K, Hertrich R, Lindner F, Rusch A, Sashko D, Schneider D, Totounferoush A, Volland D, Vollmer P and Koseomur O 2022 *Open Research Europe* **2** URL `https://doi.org/10.12688/openreseurope.14445.2`
[4] Kanna P R and Das M K 2009 *J Heat Transfer.* **131**
[5] Ramšak M 2015 *Int. J. heat mass transf.* **84** 791–799
[6] Gartling D K 1990 *Int J Numer Methods Fluids.* **11** 953–967
[7] Teruel F E and Fogliatto E 2013 *Mecánica Computacional* **32** 3265–3278
[8] Chourdakis G, Schneider D and Uekermann B 2023 *OpenFOAM® Journal* **3** 1–25 URL `https://journal.openfoam.com/index.php/ofj/article/view/88`
[9] Irons B M and Tuck R C 1969 *International Journal for Numerical Methods in Engineering* **1** 275–277

Fluid-structure interaction modeling of dry wire drawing by coupling OpenFOAM models of lubricant film and metal wire

M Vervaecke[1,*], **D Fauconnier**[2,3], **J Degroote**[1,3]
[1]Department of Electromechanical, Systems and Metal Engineering, Ghent University, Sint-Pietersnieuwstraat 41, 9000 Ghent, Belgium
[2]Department of Electromechanical, Systems and Metal Engineering, Ghent University, Technologiepark Zwijnaarde 46, 9052 Zwijnaarde, Belgium
[3]Flanders Make MIRO, Ghent, Belgium

E-mail: mathieu.vervaecke@ugent.be

Repository:https://github.com/pyfsi/coconut/tree/wire_drawing

Abstract. A steady 2D axisymmetric fluid-structure interaction model of dry wire drawing is developed to numerically investigate the interaction between the thin lubricant film and the plastically deforming steel wire. To reduce the computational cost, a layering technique is implemented in the axially moving structure. Additionally, a no-slip condition, imposed on the implemented sliding fluid-structure interaction interface, captures accurately the physics during the plastic deformation of the wire. An Arbitrary-Langrangian-Eulerian cell-centred finite volume solution methodology using pimpleFoam has been adopted to model the fluid, while a Lagrangian cell-centred finite volume solution methodology in foam-Extend executes the structural calculations. Moreover, the Python-based in-house FSI coupling code CoCoNuT performs the coupling of the flow solver and the structural solver by using the quasi- Newton IQN-ILS technique. The stresses with corresponding displacements are shown on the structure side. On the fluid side, the focus is on the behaviour of the loads of the lubricant. Additionally, the evolution of the fluid film thickness and the lubricant flow field are validated in terms of Couette and Poiseuille flow. Finally, the presented multi-physical problem shows a converged solution with a good performance of the IQN-ILS solver.

1. Introduction

Dry wire drawing is a cold work hardening process employed to gradually reduce the cross-sectional area of a wire by pulling it through a series of dies. The metal wire passes through a stationary reservoir with soap granules, from which soap is entrained in between the moving wire and a stationary die. The use of a sodium- or calcium-based soap lubricant in dry wire drawing is essential to reduce the drawing forces and the associated power. Additionally, dry wire drawing is a typical example of (plasto-) elastohydrodynamic lubrication ((P)EHL), where the surfaces of the die and wire are fully separated by a lubricant, in order to extend the service life of the die.

Optimizing the dry wire drawing process in industry is typically done via expensive and time-consuming experimental methodology based on trial-and-error. Moreover, it is experimentally not feasible to measure quantities such as pressure and film thickness in the wire-die contact zone. To give an idea of the order of magnitude for these quantities during a metal forming process, Montmitonnet *et al.* [1] estimates the maximum lubricant pressure around 2.5 GPa, while the film thickness varies between 0.5 and 2 µm. It is expected that by using a numerical model of the metal forming process, the number of experiments can be reduced. Moreover, the numerical results are obtained without interference of measurement equipment in the process.

Historically, the thin lubricant film in the PEHL contact between the die and the wire contact has usually been determined by the Reynolds equation, which was initially developed assuming an incompressible and isoviscous lubricant [2]. Later, this equation have been generalized to incorporate variable density and viscosity [3–5]. However, care must be taken that these extensions do not invalidate the assumptions in deriving the Reynolds equation. Since they are applied retrospectively, they may lead to inconsistencies [6, 7]. More recently, with the increased availability of computational power, high-fidelity CFD solvers available in OpenFOAM, have been used to more accurately capture the physical behaviour of the lubricants in narrow gaps [8, 9].

Since the shape and thickness of the lubricant film in the contact during wire drawing follows from the interaction between the hydrodynamic pressure distribution and the elasto-plastic deformation of the metal wire surface, the phenomenon is a strongly coupled fluid-structure interaction (FSI) problem. An abstract FSI problem, as illustrated in Fig. 1, consists of a structural domain Ω_s and a fluid domain Ω_f. The boundaries of both domains are indicated as Γ_s and Γ_f, respectively. The intersection of the structure and fluid boundaries is defined as the FSI interface and is denoted as $\Gamma_i = \Gamma_s \bigcap \Gamma_f$.

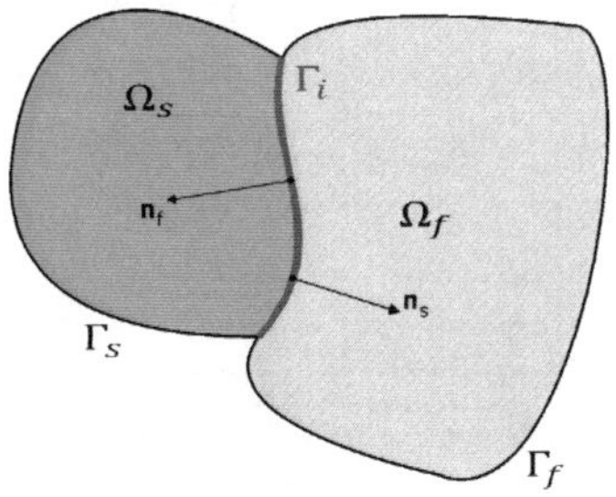

Figure 1: The structure subdomain Ω_s, the fluid subdomain Ω_f, the corresponding boundaries Γ_s and Γ_f and the fluid-structure interface Γ_i.

Several toolboxes have been developed to execute FSI simulations based on the finite volume library OpenFOAM. On one hand, Wagner *et al.* [10] present an integrated OpenFOAM membrane FSI solver for small deflections. On the other hand, solids4foam is an FSI toolbox in OpenFOAM that can handle large deflections, thanks to its extensive library of solid solvers. This toolbox is developed by Cardiff *et al.* and Tukovic *et al.* [11, 12] and has been intensively used for example in the FSI analysis of roller bearings executed by Havaej *et al.* [13, 14]. Additonaly, St-Onge *et al.* [15] present a modular multiphysical framework based on solids4foam, focussing on the implementation of formal implicit time-marching schemes to improve the stability of coupled problems.

An alternative method to execute FSI simulations with the OpenFOAM library, is to approach the OpenFOAM environment as a black box. In this case the data, calculated by the solvers, is only accessed on the fluid-structure interface. This has the advantage that dedicated and mature single-physics solvers can be used from a third-party (open source or commercial). A first example using this method is the open-source coupling library preCICE, which is extensively

IOP Conf. Series: Materials Science and Engineering **1312** (2024) 012008 doi:10.1088/1757-899X/1312/1/012008

described by Chourdakis *et al.* [16]. This library includes, besides fluid-structure interaction, also heat transfer simulations, making it capable of solving complex multi-physics problems. A second example, which uses the black box approach is CoCoNuT. The coupling code CoCoNuT, which is developed in-house at the authors' institute, is the selected FSI tool in this work and will be outlined in the remainder of this paper. CoCoNuT, abbreviation for Coupling Code for Numerical Tools, is a code under active development and is a light-weight Python package. Since CoCoNuT approaches the solver as a black box, the OpenFOAM library is only activated by this Python code. This is not to be confused with the work of Rodriguez *et al.* [17] , where an embedded Python interpreter is used directly in the OpenFOAM solvers to execute manipulation. Moreover, CoCoNuT provides different algorithms based on strongly coupled partitioned quasi-Newton methods, which stabilize and accelerate the coupling iterations [18, 19]. Additionally, this code has a fully modular approach, which is versatile and straightforward to modify.

Fig. 2 presents a schematic overview of a multi-physics problem coupled by CoCoNuT, where the structure and fluid calculations are executed by foam-Extend 4.1 and OpenFOAM 8, respectively. The exchange of data at the fluid-structure interface for each coupling iteration is executed by the communication between the so-called *solver wrappers* in CoCoNuT and the corresponding solvers. Moreover, in most cases the mesh on the fluid and structure domain is non-conformal. Consequently, an interpolation or mapping has to be executed in both directions, such that the data is imposed at the correct position on the FSI interface. Inside CoCoNuT the mappers are solver-independent and can be swapped due to the modular architecture of this coupling code. Delaissé *et al.* [20] use the FSI toolbox CoCoNuT in a numerical roller bearing analysis, where the lubricant is solved by OpenFOAM 8 and the structure is calculated by the open-source finite element library KRATOS [21]. This study has been compared with the roller bearing analysis in solids4foam executed by Havaej *et al.* [14] and presents limited discrepancies, which can be explained by the absence of the energy equation for the solid and conjugate heat transfer on the fluid-structure interface.

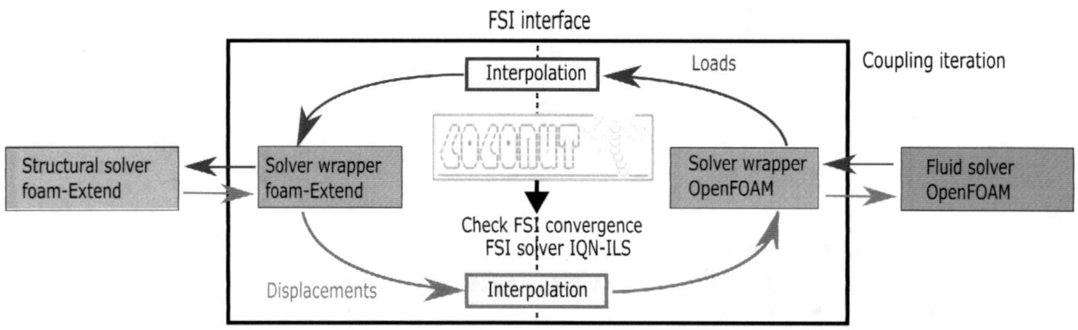

Figure 2: General overview data transfer in CoCoNuT during FSI simulation in wire drawing.

In this study an FSI solution of dry wire drawing is presented. First a description of the general principles of strongly coupled partitioned FSI method is given in section 2 , followed by a detailed outline of the data transfer between a solver in OpenFOAM 8 and its corresponding solver-wrapper in CoCoNuT. Afterwards, in section 3 the methodology to execute a 2D axisymmetric FSI simulation of dry wire drawing is defined in detail. The present model builds on a previous attempts to simulate the dry wire drawing process with FSI and a sliding FSI interface, as the wire moves axially in the lubricant domain [22, 23]. The fluid film is calculated by the pimpleFoam solver inside OpenFOAM 8 [24]. The wire deformation is obtained from the linear momentum balance and performed by a Langrangian solver in foam-Extend 4.1 [25], which is closely related to the solids4foam toolbox [11, 12]. A layering technique on the axially moving structure is implemented in order to achieve a time independent solution and a reduction in computational cost. Finally, section 4 presents the results on structure and fluid side, obtained by the numerical simulation. In this section, a grid refinement study of the lubricant film is shown. Moreover, special attention is given to the load distribution of the lubricant, which is validated in terms of Couette and Poiseuille behavior. Section 5 closes this article with a conclusion.

2. CoCoNuT: General principle

Besides to satisfying the flow and structure equations in the respective subdomains Ω_f and Ω_s subject to the appropriate boundary conditions on $\Gamma_f \backslash \Gamma_i$ and $\Gamma_s \backslash \Gamma_i$, presented in Fig. 1, the solution of a strongly coupled partitioned FSI approach must also satisfy the kinematic and dynamic condition for each time step on the FSI interface Γ_i [18]. The kinematic condition is defined as

$$\vec{v_f} = \frac{D\vec{u_s}}{Dt} \tag{1}$$

for all points on the interface, with the fluid velocity $\vec{v_f}$ and the structural displacement given by $\vec{u_s}$. Furthermore, the dynamic condition is the equality of the traction at the interface, but opposite in sign

$$\sigma_f \cdot \vec{n_f} = -\sigma_s \cdot \vec{n_s}, \tag{2}$$

with σ the stress tensor and $\vec{n}$ the unit normal vector pointed outwards from the corresponding domain (Fig. 1). The flow and structure solver are briefly written by $F(\mathbf{x})$ and $S(\mathbf{y})$, respectively. The displacements of the grid points on the interface are represented by $\mathbf{x}$, while $\mathbf{y}$ gives the loads on the fluid-structure interface. The output of a solver is denoted with a tilde.

$$\widetilde{\mathbf{y}}_{k+1} = F(\mathbf{x}_{k+1}) \tag{3}$$

$$\widetilde{\mathbf{x}}_{k+1} = S(\mathbf{y}_{k+1}) \tag{4}$$

The coupling iteration is represented by $k+1$. For each coupling iteration, the convergence criterion is checked (Fig. 2). When the FSI-convergence is not yet reached a modified displacement is imposed on the fluid domain by an improved guess of the Jacobian in the Newton-Raphson formulation of the IQN-ILS solver. This estimated Jacobian is based on the previous coupling iterations and to a certain extent on the previous time steps of the simulation [18–20]. The next time step starts when the FSI-convergence of current time step is reached. Furthermore, the residual $\mathbf{r}$ of the FSI-loop is calculated as

$$\mathbf{r}_{k+1} = \widetilde{\mathbf{x}}_{k+1} - \mathbf{x}_{k+1} \tag{5}$$

Convergence for a partitioned approach is reached when $\|\mathbf{r}\|_2 \leq \epsilon_{\mathbf{x}}$, with $\epsilon_{\mathbf{x}}$ the absolute convergence tolerance.

To perform an FSI simulation with CoCoNuT, the mandatory files and folders before the start of the calculations are presented in the green box of Fig. 3. The fluid and structure settings, boundary conditions and geometry are stored in the *Setup fluid case* and *Setup structure case*, respectively. To launch an FSI calculation in CoCoNuT, the following commands have to be executed.

IOP Conf. Series: Materials Science and Engineering **1312** (2024) 012008 doi:10.1088/1757-899X/1312/1/012008

```
./setup_structure.sh; ./setup_fluid.sh; python run_simulation.py
```

The red box in Fig. 3 illustrates the results of executing the first 2 bash scripts `setup_structure.sh` and `setup_fluid.sh`. These commands copy the corresponding setup-folders in a CSM (computational structural mechanics) and CFD (computational fluid dynamics) folder. By executing the `run_simulation.py` script, the communication between the coupling tool CoCoNuT and the solvers is activated by their corresponding *solver wrappers*. The solver wrapper is responsible to launch its solver and provides the data obtained from the coupling tool to the solver. When the solver finishes the computation during a coupling iteration, the solver wrapper receives the data and transfers it to the coupled solver (Fig. 2). Consequently, each solver wrapper is highly customized as each solver is different. However, regardless which solver is used, each solver wrapper needs the basic functions presented on the left side of Fig. 4 to accomplish all its tasks. As OpenFOAM and foam-Extend have a similar layout, only the method of communication between the OpenFOAM solver pimpleFOAM and the coupling tool CoCoNuT will be detailed in this work.

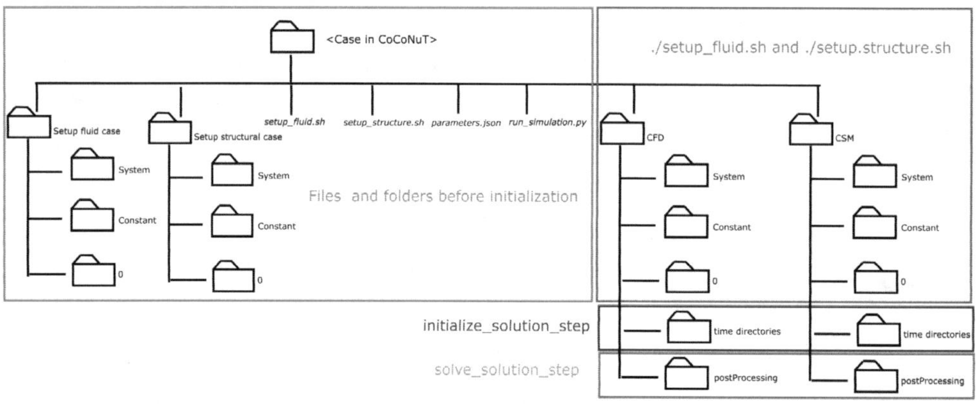

Figure 3: FSI setup in CoCoNuT with OpenFOAM as solver for both fluid and structure.

The first function activated in the solver wrapper class, is the `__init__` method (Fig. 4)[1]. The case settings in the JSON-file are read into the corresponding Python-variables, followed by the required modifications of the solver files (*e.g.controlDict*). Additionally, some checks, whether the correct modules are loaded, are executed. Finally, the variables on each interface are stored in the data structure of CoCoNuT.

The `initialize` method launches the solver, which is adapted to work with CoCoNuT, in a subprocess. Fig. 4 elucidates the most significant modification of the OpenFOAM solver to let it run in a subprocess, by putting the solver code in a infinite loop starting with `while(true)` instead of the original OpenFOAM-expression `while(runTime.run())`. This implies that CoCoNuT and not OpenFOAM controls whether the final step is reached. Moreover, the `initialize` function initializes other variables, such as time step index, physical time and coupled solver from the JSON-file.

The `initialize_solution_step` function is called at the beginning of each time step (Fig. 4).

[1] `https://github.com/pyfsi/coconut/tree/wire_drawing/coupling_components/solver_wrappers/openfoam/openfoam.py`

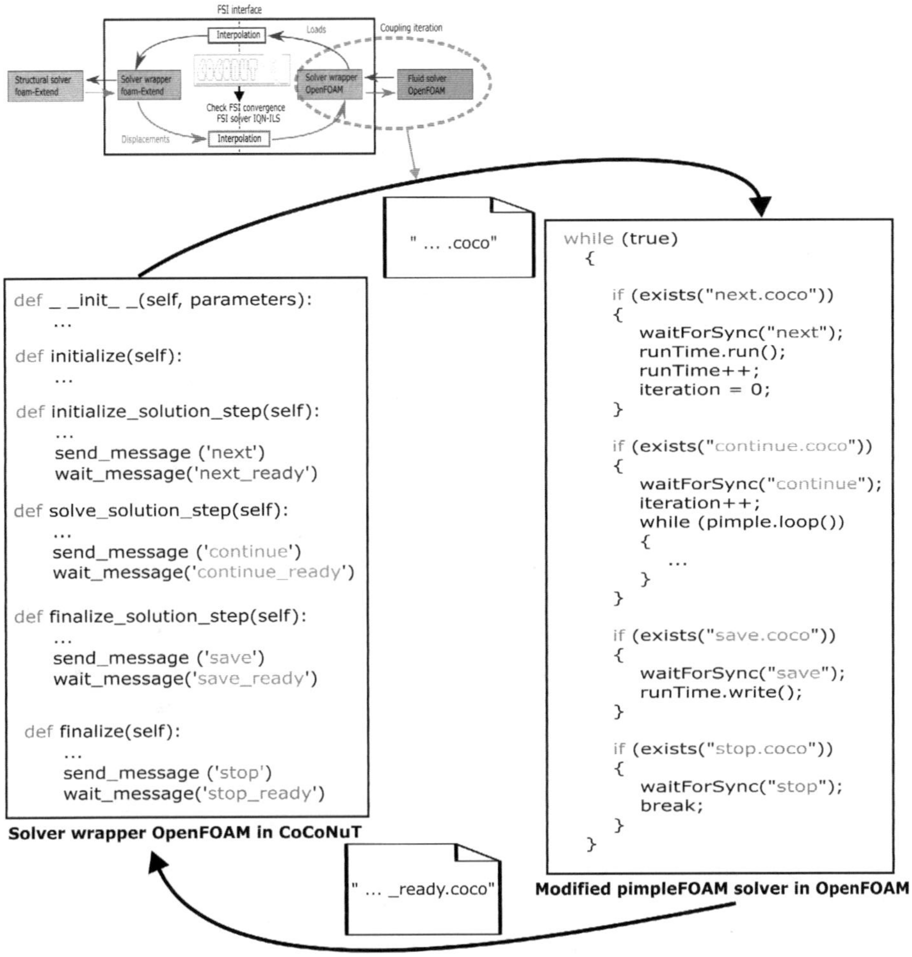

Figure 4: Communication between OpenFOAM solver wrapper and the modified pimpleFoam solver.

The time step index is increased by one and the coupling iteration index is set to zero. Additionally, inside the CFD and CSM directory a corresponding time file is created where the data of the variable fields needs to be stored, which is indicated by the blue box in Fig. 3.

The `solve_solution_step` method is called each coupling iteration. It provides input data, which is stored in an *Interface* object inside CoCoNuT (Fig. 2). This solve-solution-step function extracts the data and supplies it to the corresponding solver. Additionally, it starts the calculation and reads the output data when the solver has finished computing. At last, this method returns the data to the coupled solver as an *Interface* object as well. Moreover, the interface data for each solver is stored in the postProcessing-directory, which is constructed from the *controlDict*-file and highlighted by the orange box in Fig. 3.

The `finalize_solution_step` function checks if other flow variables have to be written in the CFD or CSM directory.

At the end of all calculations, the wrapper function `finalize` stops the solver subprocess in OpenFOAM, which was launched in the `initialize` method.

IOP Conf. Series: Materials Science and Engineering **1312** (2024) 012008 doi:10.1088/1757-899X/1312/1/012008

From Fig. 4 it can be observed that, except for `__init__` and `initialize`, each method in the solver wrapper contains a `send_message()` and `wait_message()` function. The function `send_message()` creates files with `.coco` extension. These files are used to pass messages between the Python-script and OpenFOAM. After sending a message from Python to OpenFOAM, the Python-script is paused until it receives a `..._ready.coco` message from OpenFOAM. Meanwhile, OpenFOAM executes the corresponding action and reverts into its infinite `while`-loop.

As cases can run parallel in OpenFOAM, all activated processors need to be synchronized to avoid problems while outputting results to the FSI solver. Moreover, the waitForSync.H-file is executed, which makes use of an MPI reduce command to synchronize all processors before proceeding with the rest of the code block[2].

Besides the introduction of the infinite `while`-loop in the pimpleFoam-solver, the right side in Fig. 4 presents how the communication between the OpenFOAM solver and CoCoNuT[3] is provided. First, after the OpenFOAM-solver receives the `next.coco` message, the runTime-object should be increased by one. It is important to add here the `runTime.run();` as such all the functionObjects in the controlDict-file are created and initialized to use them for storing the interface loads in the postProcessing-directory. By getting the `continue.coco` message, the governing equations are solved. Consequently, the corresponding `if`-statement contains the original solver definition. When the OpenFOAM-solver receives the `save.coco` message, the flow variables are stored in corresponding files. At last, the `while(true)`-loop ends by activating a `break`-statement after sending `stop.coco` to the OpenFOAM solver.

3. Methodology: FSI calculation wire drawing

This section presents the geometry, boundary conditions and settings of simulating a dry wire drawing process with a 2D axisymmetric model. Moreover, emphasis is placed on the algorithm to solve such a complex multi-physics problem as dry wire drawing. The quasi-Newton IQN-ILS algorithm inside CoCoNuT performs the coupling of the flow solver and structural solver [18–20]. Furthermore, the approach of the fluid and structural simulations are outlined. In this ongoing study a high constant value of the viscosity is assumed to avoid die-wire contact during the simulation. For this reason, the pimpleFoam solver in OpenFOAM 8 currently computes the lubricant film [24]. The piezo-viscous and shear-thinning behavior of the lubricant as well as the temperature effects will be implemented and investigated in a later phase. Special attention is given to the implementation of the sliding FSI interface and the no-slip condition at the FSI interface to capture accurately the physics during the cross section reduction of the wire. At last, the implementation of adding and removing cell layers on the structure side is briefly described.

3.1. Geometry, boundary conditions and settings

In the presented axisymmetric model, the stationary soap reservoir and the transition from soap granules to a soap film are not taken into account. Consequently, the soap lubricant is approximated as a fluid from the onset. Atmospheric pressure is imposed at the inlet and outlet of the lubricant domain.

A difficulty in modelling dry wire drawing is the initialization of the simulation. Therefore, a procedure has been developed with an initially undeformed wire and an artificially enlarged die. Fig. 5 illustrates the initial condition of the wire drawing simulation with everything at

[2] `https://github.com/pyfsi/coconut/tree/wire_drawing/coupling_components/solver_wrappers/openfoam/waitForSync.H`

[3] `https://github.com/pyfsi/coconut/tree/wire_drawing/coupling_components/solver_wrappers/openfoam/v8/coconut_pimpleFoam`

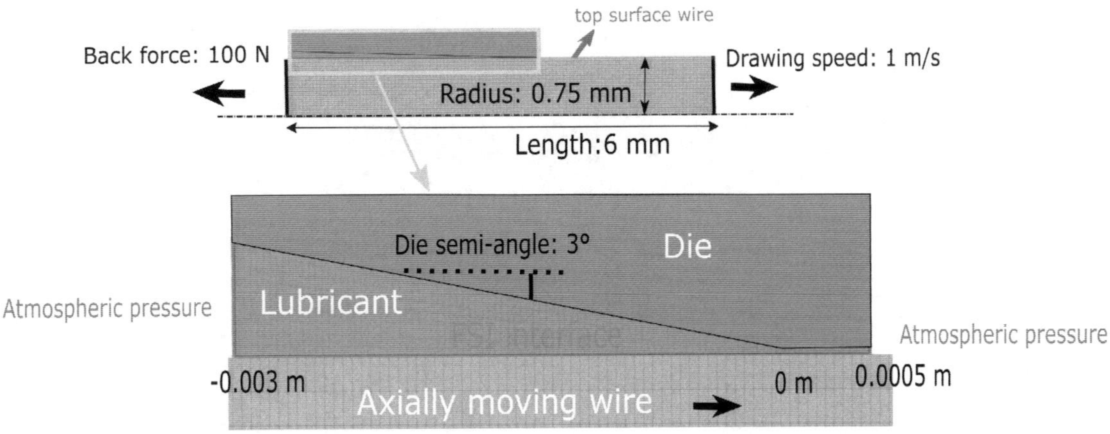

Figure 5: Axisymmetric wire-die configuration at initial condition, close up is 5 times scaled in radial direction.

rest. It consists of an originally undeformed wire, which passes through a die with an initially enlarged diameter. As shown in Fig. 5 the space between the wire and the die is represented by the lubricant having a predefined film thickness. During the simulation, the die diameter decreases gradually until it reaches its real diameter, while the wire passes through the die and deforms until a new wire radius is reached. Fig. 6 represents a steady-state result, with correct die diameter. Consequently, the die is approximated as a body with prescribed deformation [26], while the elastoplastic deformation of the wire is calculated.

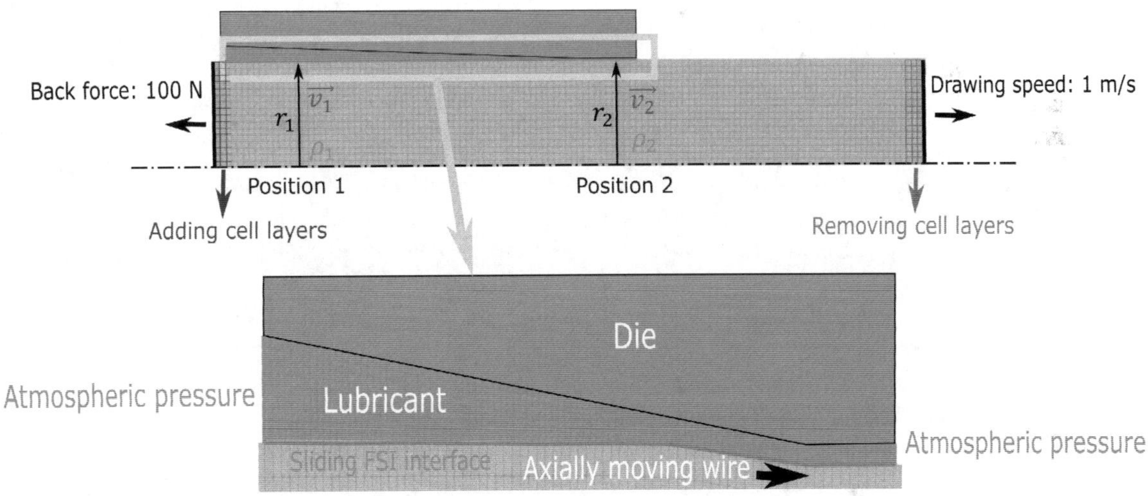

Figure 6: Axisymmetric wire-die configuration at steady state with wire radius r, wire drawing velocity $\vec{v}$ and structural density ρ at position 1 and 2. Position 1 represents the structural material properties before the wire deformation. The close up is 5 times scaled in radial direction.

A constant axial drawing velocity of 1 m/s is applied on the right end of the wire. The axial velocity of the lubricant is coupled with the axial top surface velocity of the wire (Fig. 5), which preserves the no-slip condition between the wire and the boundary of the lubricant contacting the wire. Furthermore, a back force of 100 N is applied on the left end of the wire.

Table 1: Mechanical properties of the wire

Young's modulus	E	177 GPa
Density	ρ	7833 kg/m^3
Poisson coefficient	v	0.3
Initial yield stress	σ_y	1.25 GPa
Hardening coefficient		0.3

Table 2: Numerical settings

Number of cells wire	1952
Number of cells lubricant	900
Number of faces wire on FSI interface	76
Number of faces lubricant on FSI interface	100
Absolute tolerance FSI coupling	0.005 µm
Time step (Δt)	0.00005 s
Total number of time steps (n)	2000
Time steps before shrinking die diameter	20
Time steps during shrinking die diameter	1800

The mechanical properties of the wire are specified in Table 1. Following [22], the wire diameter is 1.5 mm and the die angle is 6°. Additionally, a constant lubricant viscosity of 350 Pas is applied. Table 2 gives the settings concerning the numerical calculations. The discretization of the structural domain is based on the mesh of the wire presented in [3], where the domain in radial direction is divided in layers of 16 cells. Additionally, the fluid domain is meshed with 900 cells to solve current multi-physical problem, which is based on a mesh study presented in section 4. It is an unsteady simulation which is executed until steady conditions are reached. The case is solved for time $t = [0\text{s}, 0.1\text{s}]$. The determination of the time step Δt is based on the application of the layer technique, outlined at the end of this section. During the first 20 time steps the die has the artificially enlarged diameter to avoid numerical instabilities, caused by the acceleration of the wire from 0 to 1 m/s. After these 20 time steps the die diameter starts to shrink each time step until reaching the actual die diameter at time step 1820, where a maximum cross-section reduction of 10% of the wire is achieved. The total simulation takes 2000 time steps (n).

3.2. Algorithm FSI solution procedure

Algorithm 1 presents the solution procedure for an FSI wire drawing calculation with a strongly coupled partitioned approach. To obtain sufficient quality of the FSI interface, mapping is required. The mapping blocks M_{FS} and M_{SF} execute the required transfer of data from the fluid domain to the structure domain and the other way around, respectively. Each mapping block contains an axisymmetric to 2D and 2D to axisymmetric mapper, with a linear interpolation of both grids in the intermediate 2D space[4].

Line 8 in Algorithm 1 states that the quasi-Newton IQN-ILS algorithm is used as FSI-solver. Furthermore the absolute convergence tolerance $\epsilon_{\mathbf{x}}$ is equivalent to an accuracy of 0.005 µm on the radial displacement of the wire.

[4] `https://github.com/pyfsi/coconut/tree/wire_drawing/coupling_components/mappers`

Algorithm 1 Solution procedure FSI calculation wire drawing

for all time steps **do**
  k = 0
  $\tilde{\mathbf{y}}_0 = F(\mathbf{x}_0)$
  $\mathbf{y}_0 = M_{FS}(\tilde{\mathbf{y}}_0)$
  $\tilde{\mathbf{x}}_0 = S(\mathbf{y}_0)$
  $\mathbf{r}_0 = M_{SF}(\tilde{\mathbf{x}}_0) - \mathbf{x}_0$
  while $\|\mathbf{r}_k\|_2 > \epsilon_{\mathbf{x}}$ **do**
    $\mathbf{x}_{k+1} = \mathbf{x}_k + \Delta\mathbf{x}_k^{IQN}$
    $\tilde{\mathbf{y}}_{k+1} = F(\mathbf{x}_{k+1})$
    $\mathbf{y}_{k+1} = M_{FS}(\tilde{\mathbf{y}}_{k+1})$
    $\tilde{\mathbf{x}}_{k+1} = S(\mathbf{y}_{k+1})$
    $\mathbf{r}_{k+1} = M_{SF}(\tilde{\mathbf{x}}_{k+1}) - \mathbf{x}_{k+1}$
    $k++$
  end while
  Update yield stress structure
  Update mesh structure
end for

3.3. Fluid simulations

During the simulations the mesh of the fluid film will deform and hence the Arbitrary-Lagrangian-Eulerian (ALE) formulation is used. The conservation of mass and the momentum equations are given by

$$\frac{\delta}{\delta t}\int_{V_f} \rho_f \,\mathrm{d}V_f + \oint_{S_f} \rho_f(\vec{v_f} - \vec{v_{fg}}) \cdot \vec{n_f}\,\mathrm{d}S_f = 0 \tag{6}$$

$$\frac{\delta}{\delta t}\int_{V_f} \rho_f \vec{v_f}\,\mathrm{d}V_f + \oint_{S_f} (\rho_f \vec{v_f}(\vec{v_f} - \vec{v_{fg}})) \cdot \vec{n_f}\,\mathrm{d}S_f = \oint_{S_f} \sigma_f \cdot \vec{n_f}\,\mathrm{d}S_f \tag{7}$$

where V_f is the volume and S_f the surface of the cell. The flow velocity is represented by $\vec{v_f}$, $\vec{v_{fg}}$ is the grid velocity, ρ_f the fluid density and t the time. To compute the grid velocity the so-called *space conservation law (SCL)* [27] is applied.

$$\frac{\delta}{\delta t}\int_{V_f} \mathrm{d}V_f - \oint_{S_f} \vec{v_{fg}} \cdot \vec{n_f}\,\mathrm{d}S_f = 0 \tag{8}$$

In this case the fluid is approximated as a Newtonian fluid, so the stress tensor is defined as $\sigma_f = -p\mathbf{I} + 2\mu_f D$, where p is the pressure, μ_f the constant fluid viscosity and $D = \frac{1}{2}[\nabla\vec{v_f} + (\nabla\vec{v_f})^T]$ the rate of strain tensor. The first term on the left-hand side of Eqn. 7 is discretized in time with an implicit second order scheme, which guarantees accuracy with a CFL *(Courant-Friedrichs-Lewy)* number larger than 1 in combination with the Pimple-loop iterations [24,28]. The Peclet number with 1e-8 order of magnitude is significantly lower than 1, which allows to apply a second order central discretization scheme on the convection term (second term left-hand side of Eqn. 7), which gives a higher accuracy [29,30]. The term on the right-hand side of Eqn. 7 is discretized with a second order central scheme as well.

3.4. Structural simulations

Considering a body of volume V_s limited by an area S_s, the governing equation is given by the force balance and is denoted as

$$\frac{\delta}{\delta t}\int_{V_s} \rho_s \vec{v_s}\,\mathrm{d}V_s = \oint_{S_s} \sigma_s \cdot \vec{n_s}\,\mathrm{d}S_s \tag{9}$$

where ρ_s presents the solid density, $\vec{v_s}$ the velocity vector and σ_s the Cauchy tensor of the structure. Moreover, the relation between the stress tensor σ_s and the strain is given by the constitutive equation of the material and is defined in detail in [25]. Using the Nanson's relations, the deformed surface S_s can be associated to the initial area S_{0s} by $S_s = J\mathbf{F}^{-T} \cdot S_{0s}$, where $\mathbf{F}$ is the deformation gradient and J the Jacobian of the deformation gradient [25,31]. Moreover, Eqn. 9 can be reformulated with regard to the initial undeformed configuration, indicated by subscript 0:

$$\int_{V_{0s}} \frac{\delta}{\delta t}\rho_{0s}\vec{v_s}\,\mathrm{d}V_{0s} = \oint_{S_{0s}} (J\mathbf{F}^{-T} \cdot \vec{n_{0s}}) \cdot \sigma_s\,\mathrm{d}S_{0s} \tag{10}$$

or in terms of the updated Lagrangian formulation:

$$\int_{V_s} \frac{\delta}{\delta t}\rho_s\vec{v_s}\,\mathrm{d}V_s = \oint_{S_s} (j\mathbf{f}^{-T} \cdot \vec{n_s}) \cdot \sigma_s\,\mathrm{d}S_s \tag{11}$$

where $\mathbf{f}$ is the relative deformation gradient and relates the current deformation state with the previous one. Moreover, j is the Jacobian of the relative deformation gradient.

To enable a segregated solution procedure, the surface force (first term right hand side Eqn. 11) is divided into implicit and explicit components. This procedure is defined as [25]:

$$\begin{aligned}\int_{V_s} \frac{\delta}{\delta t}(\rho_s \frac{\delta(\vec{u_s}^{m-1} + \Delta\vec{u_s}^{e})}{\delta t})\,\mathrm{d}V_s = \oint_{S_s} ((\frac{4}{3}\mu_s + K)\vec{n_s} \cdot \nabla(\Delta\vec{u_s}))^e\,\mathrm{d}S_s + \oint_{S_s} ((j\mathbf{f}^{-T} \cdot \vec{n_s})\sigma_s)^{e-1}\,\mathrm{d}S_s \\ - \oint_{S_s} ((\frac{4}{3}\mu_s + K)\vec{n_s} \cdot \nabla(\Delta\vec{u_s}))^{e-1}\,\mathrm{d}S_s\end{aligned} \tag{12}$$

where $\vec{u_s}$ is the displacement vector of the structure, μ_s the shear modulus and K the bulk modulus. Furthermore, the relative deformation gradient is described in terms of the displacement increment as $\mathbf{f} = \mathbf{I} + \nabla(\Delta\vec{u_s})^T$. The displacement increment is given by $\Delta\vec{u_s} = \vec{u_s}^{m} - \vec{u_s}^{m-1}$. The superscripts m and $m-1$ indicate values of current and previous time step and the superscripts e and $e-1$ indicate values of the current and previous iteration of the structural solver, respectively. The component containing $\Delta\vec{u_s}^{e}$ of the first term at the left hand side of Eqn. 12 is discretised using a first-order accurate in time fully implicit backward differencing scheme, while the term containing $\vec{u_s}^{m-1}$ is calculated in an entirely explicit manner. Consequently, the latter contributes only to the source vector of the resulting linear system of equations. The first term at the right hand side of Eqn. 12 is discretised using central differencing with over-relaxed non-orthogonal correction. The second and third terms on the right-hand side of Eqn. 12 are discretised using central differencing as well. Recall that the first term and the third term of Eqn. 12 do not exactly cancel out, because the first term is treated implicitly, while the third term is approximated explicitly. The difference of both terms results in additional numerical diffusion, which tends to eliminate erroneous oscillations. The complete and detailed formulation of the structural solver, with the numerical implementation is described in [25,32].

Originally, the structural solver was created for simulation of metal forming processes without fluid interaction [25]. During the wire drawing process, work hardening occurs. Consequently, to calculate a correct true stress, the yield stress and the mesh deformation have to be updated each time step. In an FSI simulation the yield stress field and structural mesh are only updated after the coupling iteration has reached the convergence criterion, which is elucidated in Algorithm 1 at line 15 and 16. Algorithm 2 presents the structural solver solution procedure for each coupling iteration. Convergence is reached when $\mathbf{r}_s \leq \zeta$, with ζ the convergence tolerance equal to 1e-9. The scalar $\mathbf{r}_s$ is calculated by

$$\mathbf{r}_s = \frac{\|\Delta \vec{u_s}^{e} - \Delta \vec{u_s}^{e-1}\|_\infty}{\|\Delta \vec{u_s}^{e}\|_\infty} \tag{13}$$

which is a relative criterion based on the change of the displacement of the structure.

Algorithm 2 Solution procedure structural solver

for each FSI-coupling iteration **do**
  while ζ is not reached **do**
    Momentum equation: assemble and solve in terms of $\Delta\vec{u_s}$ [25]
    Calculate kinematics and stress [25]
    Calculate $\mathbf{r}_s$
  end while
  Update density and equivalent Cauchy stress
  Interpolate cell-centre displacements to the vertices
end for

3.5. Sliding FSI interface

During the simulation, the lubricant domain Ω_f stays fixed in axial direction, while the wire domain Ω_s moves along the axis (Fig. 1, Fig. 5 and Fig. 6). This creates a large translation of the structural mesh with respect to the fluid grid in axial direction. Consequently, a sliding FSI interface approach is applied [22]. As the wire moves through the die, the hydrodynamic pressure and wall shear stresses obtained from the fluid solver must be imposed on the correct position on that part of the wire located in the die. Moreover, only the grid velocity normal to the interface has to be the same on both sides of the interfaces [19]. This allows slip of the structure grid with respect to the fluid grid, even though the boundary at the FSI interface has a no-slip condition [19]. Consequently, the intersection Γ_i has to be updated each time step (Fig. 1).

In order to deal with the sliding FSI interface, the boundary condition *timeVaryingMapped-SolidTraction*[5] is developed on the structural side. The purpose of this boundary condition is to impose the loads, obtained from the FSI interface, on the correct position at the current time step m on the axially moving wire. This boundary condition is based on the idea of the timeVaryingMappedFixedValue boundary condition, which is implemented in the foam-Extend package, and the solidTraction boundary condition [11, 12, 33]. As presented in Fig. 7, during the initialization of the FSI simulation, the structural solver wrapper creates inside the *constant* folder a *boundaryData\boundaryPatch* directory named after the patch referring to the boundary at the FSI-interface. This folder carries n time step directories and a single point file. Moreover, each time step directory contains a pressure and traction file initialized with zero-entries corresponding to the number of vertices in the point file. To create the point file, the definition of the axisymmetric condition inside OpenFOAM has to be taken into account.

[5] `https://github.com/pyfsi/coconut/tree/wire_drawing/coupling_components/solver_wrappers/openfoam_extend/v41/coconut_src/boundaryConditions`

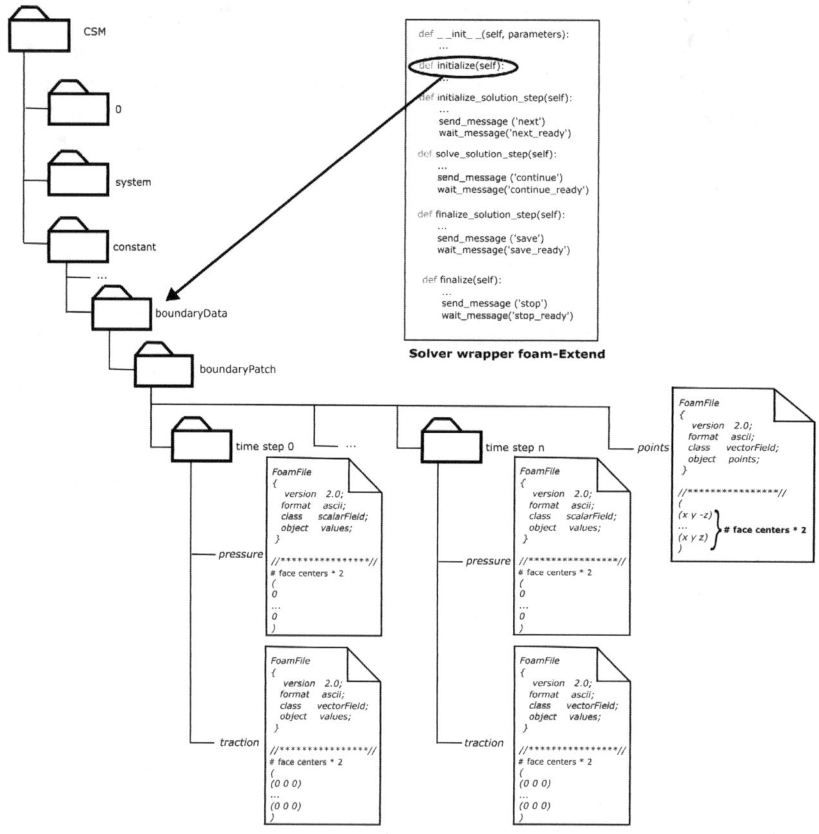

Figure 7: Schematic overview of initializing the timeVaryingMappedSolidTraction boundary condition for n time steps by creating a point file and load files with zero entries for each time step.

As presented in Fig. 8, the axisymmetric or wedge condition in OpenFOAM contains a mesh with one cell across the circumferential direction, which reduces the number of cells to two dimensions in the radial and axial directions. The wedge boundary condition is applied to the two sloping side patches, which are indicated in Fig. 8 [28]. Moreover, a wedge angle α of 5° is defined inside the OpenFOAM algorithm. To obtain the coordinates in the point file, the face centre coordinates of the top wire surface boundary (Fig. 5) are rotated by an angle of $\beta/2$ degrees in positive and negative sense of the circumferential direction (Fig. 7), represented by the blue and red dots in Fig. 8, respectively. The angle β, which is in this case 2.5°, can be arbitrarily chosen as the values of the calculated loads imposed on the FSI interface do not change in circumferential direction due to the axisymmetric definition of this case.

By activating the *solve_solution_step* function, Fig. 9 illustrates that the mapped loads obtained from the fluid solver are stored in the corresponding pressure and traction files in the *boundaryPatch* directory for the current time step m. This implies the $timeVaryingMappedSolidTraction$ boundary condition is created such that the pressure and traction files are updated each coupling iteration k. Additionally, the pressure and traction, imposed on the face centres of the top surface of the axially moving wire (Fig. 5), are obtained by a weighted interpolation of the updated loads located on the coordinates in the point file.

To close the complete sliding FSI interface implementation, only a limited amount of data has

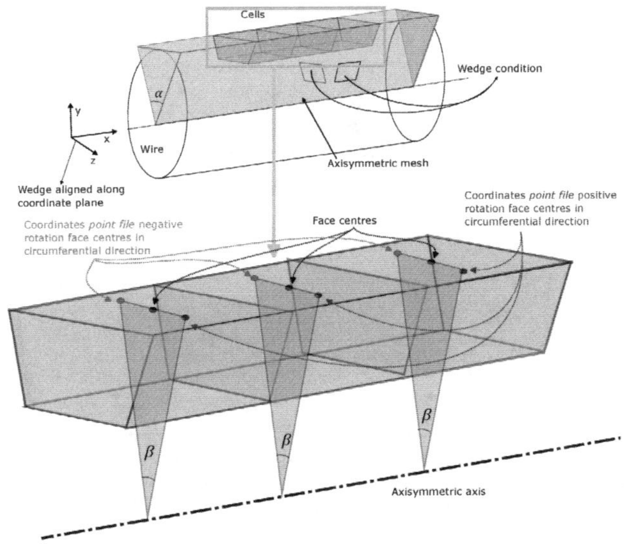

Figure 8: Axisymmetric or wedge condition in OpenFOAM/foam-Extend and rotation of the face centres in positive and negative circumferential direction.

to be transferred from structure to fluid domain. This data contains exclusively the radial displacements of the wire where it is between the left and right end of the die (Fig. 6 and Fig. 9). Only these radial displacements are used to calculate the coupling residual.

3.6. No-slip condition

In the previous subsection it has been stated that to achieve a sliding interface it is sufficient that only the grid velocity normal to the interface has to be the same on both sides of the interface. However, the axial velocity of the wire has to be imposed on the FSI interface in order to capture accurately the physics as a result of the cross-section reduction of the wire during the drawing process. This implies that the normal and the tangential material velocity have to be identical for the lubricant and wire at the interface. The first condition is satisfied by the sliding FSI interface. The second condition is achieved by a transfer of the wire's axial velocity to the fluid boundary contacting the wire in order to create a no-slip condition.

To impose the axial velocity of the wire on the lubricant boundary contacting the FSI interface a new boundary condition *timeVaryingMappedCoupledVelocity*[6] is created on fluid side. A schematic overview of applying this boundary condition is presented by Fig. 10. It can be observed that this boundary condition is based on a similar approach as the previously proposed condition *timeVaryingMappedSolidTraction*, which implies the initialization of *timeVaryingMappedCoupledVelocity* will take place in the same way for n time steps (Fig. 7). Moreover, this boundary condition updates the axial velocity of the wire each coupling iteration k for the current time step m and imposes it on the correct face centre of the lubricant according to a weighted interpolation of the axial velocity located at the coordinates in the point file. The transfers of the displacements are disregarded in Fig. 10 as they have no significant role in applying *timeVaryingMappedCoupledVelocity*. Additionally, it is important to report that using *timeVaryingMappedCoupledVelocity* has significant consequences when calculating the coupling residual. Working with this boundary condition implies, besides having converged

[6] `https://github.com/pyfsi/coconut/tree/wire_drawing/coupling_components/solver_wrappers/openfoam/v8/coconut_src/boundaryConditions`

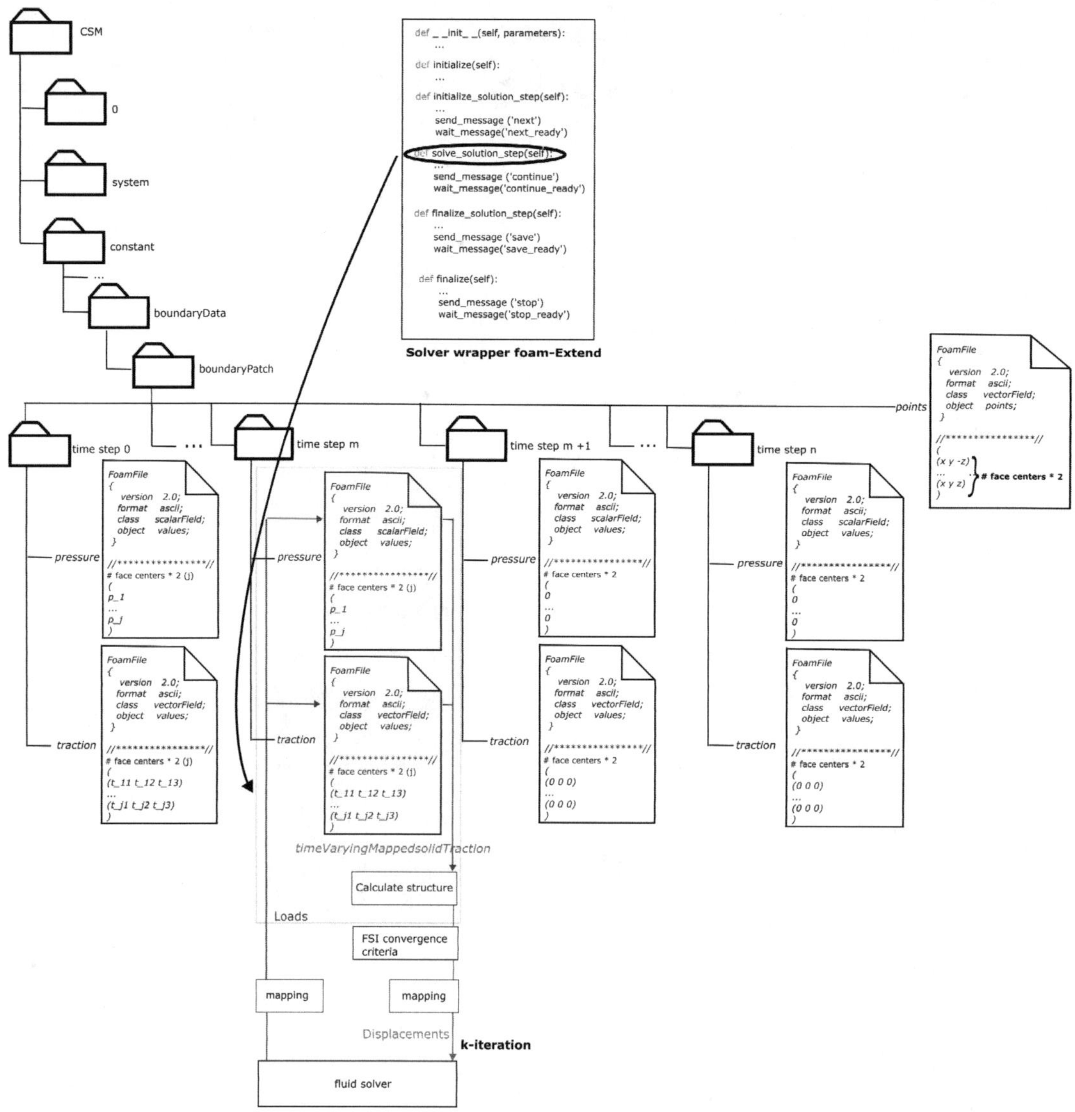

Figure 9: Schematic overview of applying the *timeVaryingMappedSolidTraction* boundary condition during the simulation by refreshing each coupling iteration k the load files for the current time step m.

results for the radial displacements, the convergence of the axial velocity has also to be taken into account in order to obtain a completely converged calculation. This can give problems as the axial velocity in this case is 5 orders of magnitude higher than the radial displacement of the wire. To limit this effect, the axial velocity of the wire has to be divided by 1e5 before transferring the data to the fluid side. Obviously, after the operations in the coupling algorithm but before imposing the axial velocity on the lubricant boundary contacting the structure, the axial velocity has to be multiplied by 1e5.

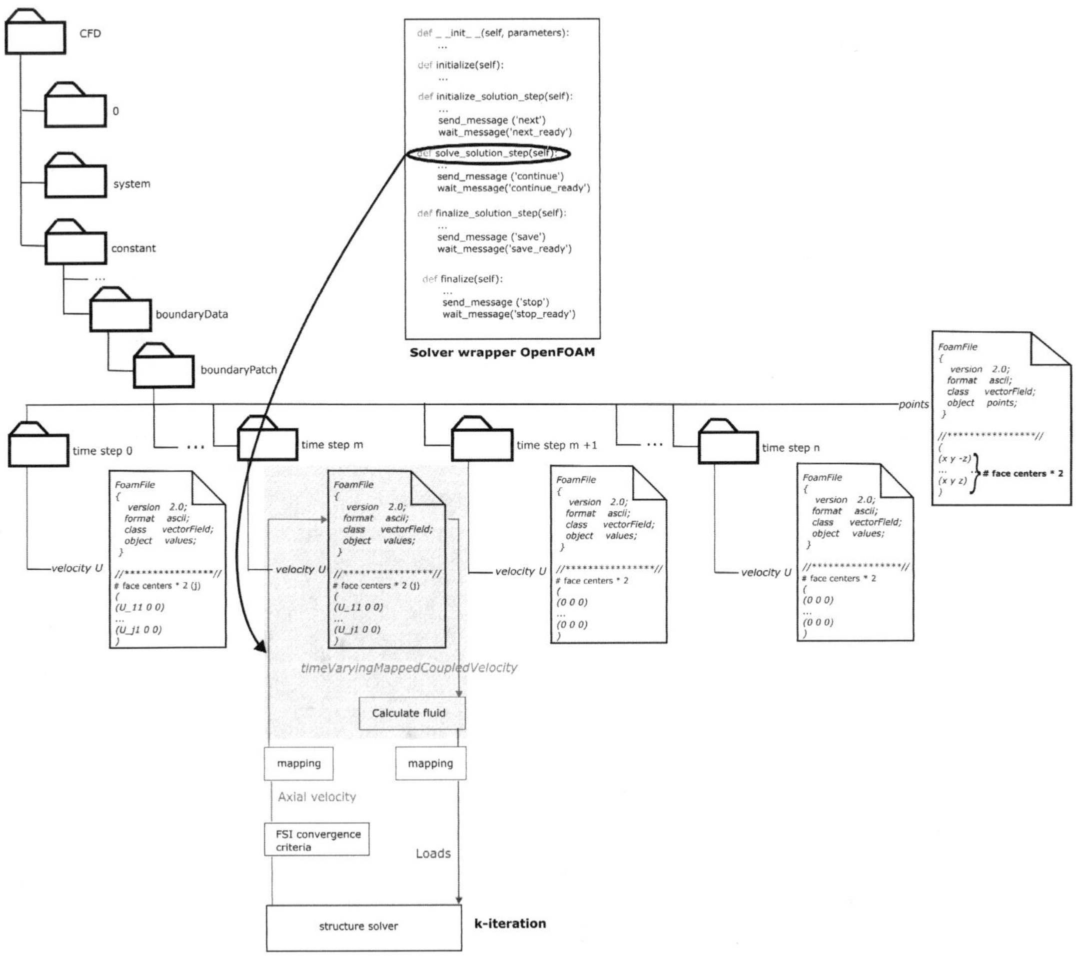

Figure 10: Schematic overview of applying the *timeVaryingMappedCoupledVelocity* boundary condition during the simulation by refreshing each coupling iteration k the velocity files for the corresponding time step m.

3.7. The layering technique

In order to ensuring a lower computational cost, a layering technique is introduced such that the deformations of the wire are only calculated in the zone of interest [34]. As illustrated in Fig. 6, this layering technique involves the removal of the deformed layer of cells on the right side of the structural domain, whereas a new layer of cells is added to the undeformed wire on the left side of the structural domain. However, combining the presented layering technique with the aforementioned no-slip condition initially caused oscillations of the radial displacement of the moving structure and thus of the velocity along the wire. Consequently, this prevented a converged solution of the FSI method.

To understand the origin of this oscillating behaviour, the wire drawing model presented in Fig. 6 has to be considered. The structural mass equation at steady state is given by

$$\rho_1 r_1^2 \vec{v}_1 = \rho_2 r_2^2 \vec{v}_2 \tag{14}$$

where ρ represent the wire density, $\vec{v}$ the speed of the wire and r the radius of the wire. The left side of Eqn. 14 is represented by the values of *position 1* in Fig. 6, while *position 2* of Fig. 6

displays the values of the right side of Eqn. 14. According to Table 1 and Fig. 6, ρ_1, $\vec{v}_2$ and r_1 have constant values during the simulation. As deformation of the metal structure occurs during the wire drawing process, $\vec{v}_1 \neq \vec{v}_2$ (Eqn. 14). Rearranging the terms of Eqn. 14 with velocity $\vec{v}_1$, which represents the wire speed before deformation, to be solved in the left term results is

$$\vec{v}_1 = \frac{\rho_2}{\rho_1} \frac{r_2^2}{r_1^2} \vec{v}_2 \tag{15}$$

According to Eqn. 15 the inlet velocity $\vec{v}_1$ depends on the ratio between the densities and the square of the radii multiplied by $\vec{v}_2$. Fig. 11 outlines the radial displacement of two consecutive time steps m and $m+1$, exposed to a constant pressure distribution. It illustrates that few data points across the width of the lubrication pressure peak are available to impose the load on the moving wire mesh. Consequently, small undesirable differences in radial direction and density occur between two consecutive time steps. This influences the velocity $\vec{v}_1$ (Eqn. 15) and causes small oscillations of the velocity profile at the inlet of the wire during the simulation (Fig. 6), which are detrimental for the convergence of the FSI algorithm.

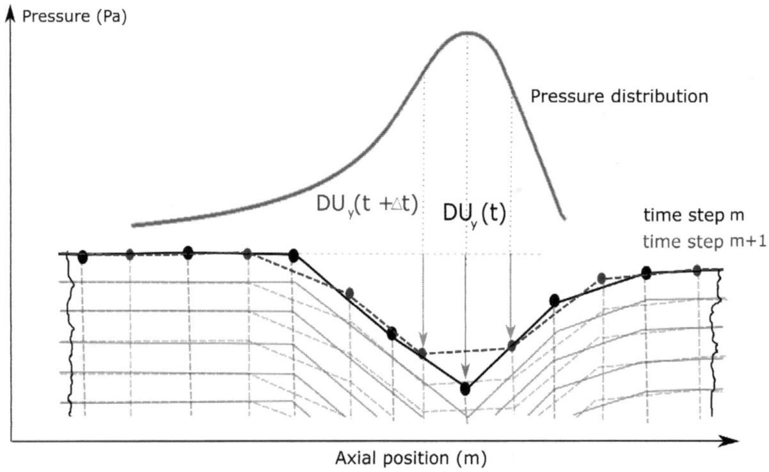

Figure 11: Radial displacement of consecutive time steps *m* and *m+1* subject to a constant load in a discretized system if the layer addition is not synchronized with the time step.

The most obvious solution is to refine the mesh. However, the computational cost increases significantly by this approach. Moreover, the amplitude of the oscillations is reduced, but they do not disappear. The chosen solution is adding a layer of cells at the left side of the structural domain each time step. This means the location of the cell in current time step is the same as the position of the preceding cell in previous time step of the axially moving structure. Consequently, each cell experiences a similar load as the preceding cells on that position (Fig. 11). This gives a similar radial displacement and density for each time step, which stabilizes the velocity $\vec{v}_1$ (Eqn. 15). Moreover, the oscillating behaviour of the radial displacement disappears and a converged FSI solution can be achieved. This implies for the wire drawing case that the time step magnitude Δt is 5e-5 s (Table 2), which is determined by the drawing speed (Fig. 5) and the layer thickness of the structural mesh [3].

4. Results

The results section first presents a grid refinement study of the FSI simulation, followed by the physical parameters at the steady state condition. Additionally an analysis of the velocity field of the lubricant in terms of Couette and Poiseuille flow is outlined. Finally, the numerical performance is presented[7].

4.1. Grid refinement study

In order to have a converged CFD result with an uncertainty estimation, based on grid refinement, of the solutions presented in current section a mesh convergence study was performed for the fluid film calculations with respect to the force exerted on the boundary of the fluid film contacting the FSI interface (Fig. 5). Table 3 and 4 present 5 CFD calculations with a mesh refinement in axial and radial direction. These simulations are executed without an FSI interaction. Consequently, the FSI interface in the CFD convergence study is defined as a rigid body boundary. Each CFD calculation is executed with a time step size Δt of 5e-5 s, which is determined by the axial layer thickness of the cells of the wire and the drawing velocity to avoid oscillating behaviour of the radial displacement during the FSI calculation. Moreover, a mesh refinement study of the wire was performed in [3] and it shows an uncertainty $U_{\phi(CSM)}$ of 0.52% of the wire mesh used in this case.

Table 3 and 4 each show 5 converged results. Furthermore, a significant difference in the resulting force can be observed depending on the mesh refinement direction, which is the main reason to execute a grid convergence study in axial and radial direction.

Table 3: Test matrix results CFD convergence study axial mesh refinement. Each simulation contains 9 layers in radial direction.

No	total cells	max CFL	F(N)
1	540	0.87	669.878
2	630	1.04	670.189
3	720	1.15	670.373
4	810	1.30	670.439
5	900	1.44	670.678

Table 4: Test matrix results CFD convergence study radial mesh refinement. Each simulation contains 100 layers in axial direction.

No	total cells	max CFL	F(N)
1	500	1.37	655.473
2	600	1.40	661.972
3	700	1.42	666.041
4	800	1.43	668.759
5	900	1.44	670.678

The freely available code developed by *Marin* [35] was used to perform the uncertainty and error estimation based on grid refinement for the fluid film. The code is based on the procedures suggested by [36]. The output of the numerical uncertainty of the fluid film based on grid refinement in the axial and radial direction are presented in Table 5. ϕ_0 represents the extrapolated exact solution, ϕ_1 denotes the finest level solution, U_ϕ displays the estimated uncertainty and p presents the achieved order in space. According to [36], an estimate of the uncertainty can only be obtained if $p > 0$, which is satisfied for the axial and radial direction. Additionally, it can be observed in Table 5 that the estimated uncertainty of the radial mesh refinement study is significantly higher than the estimated uncertainty of the axial mesh study. An important remark is the application of a safety factor of 1.25 on the estimated uncertainty. The limiting factor in achieving grid convergence and reducing uncertainty for fluid film calculations is convergence problems when the die reaches its final position during the FSI calculation. This FSI divergence occurs at 10 layers in the radial direction of the lubricant film and requires deeper investigation to understand the root of the problem.

[7] `https://github.com/pyfsi/coconut/tree/wire_drawing/examples/wire_drawing`

Table 5: Numerical uncertainty of the fluid film case based on grid refinement in axial and radial direction.

Item	ϕ_0	ϕ_1	U_ϕ	p
$F[N]_{ax}$	671.789	670.678	0.23%	1.95
$F[N]_{rad}$	678.879	670.678	1.53%	1.78

The numerical uncertainty of the fluid film based on grid refinement for the mesh containing 900 cells (100 layers axial direction, 9 layers radial direction) is obtained from Eqn. 16.

$$U_{\phi(CFD)} = \sqrt{U^2_{\phi(ax)} + U^2_{\phi(rad)}} = 1.55\% \quad (16)$$

The total uncertainty of the grid for the FSI calculation in this work is thus calculated in Eqn. 17

$$U_{\phi(total)} = \sqrt{U^2_{\phi(CFD)} + U^2_{\phi(CSM)}} = 1.63\% \quad (17)$$

4.2. Results physical parameters at the steady state condition of dry wire drawing modelling

Fig. 12 illustrates the speed at the top surface of the wire and the velocity imposed on the boundary of the lubricant contacting the wire. A perfect fit of the velocity of the wire and the speed of the lubricant at the FSI-interface is presented, which indicates a no-slip condition and thus a correct implementation of the *timeVaryingMappedCoupledVelocity* boundary condition. Additionally, the conservation of mass effect on the wire velocity is clearly illustrated in Fig. 12, by the acceleration of the wire speed during its deformation.

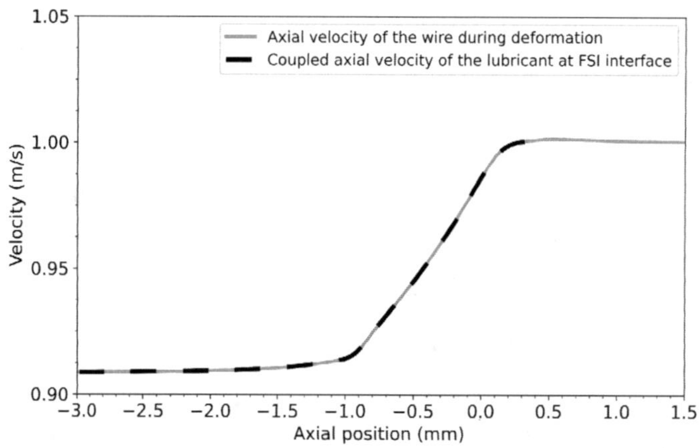

Figure 12: Velocity profile of the wire and corresponding speed imposed on the lubricant boundary contacting the wire.

Fig. 13 presents on the left side the equivalent Cauchy stress evolution of the wire during the metal deformation. The equivalent Cauchy stress is a scalar quantity representing the magnitude of stress and is in this case based on the J2 plasticity [25, 37, 38]. Moreover, Fig. 13 presents on the right side the equivalent Cauchy stress evolution at the top surface of the wire and the corresponding yield stress. Additionally, Fig. 13 illustrates the plastic deformation of the wire, while it is pulled through the die. The steep increase of the equivalent Cauchy stress is limited by the yield stress, which is the point where the wire starts to deform plastically. Moreover, the yield stress and equivalent Cauchy stress increase slightly and work hardening on the wire occurs. When the wire leaves the die, the equivalent Cauchy stress relaxes and the yield stress stays constant. Furthermore, the equivalent Cauchy stress of the wire, presented on the left of Fig. 13, shows a similar behaviour with the stress on the wire simulated in [3].

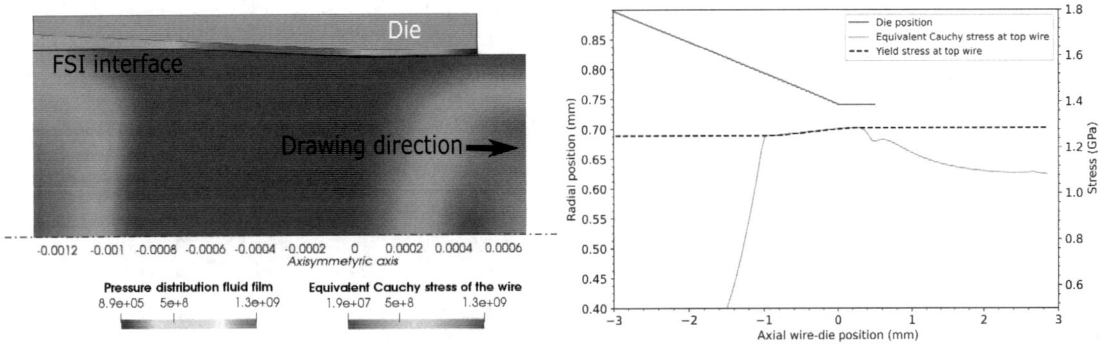

Figure 13: Left: Equivalent Cauchy stress wire and pressure distribution lubricant; Right: equivalent Cauchy stress and yield stress of structure at the FSI interface.

Fig. 14 demonstrates the radial displacement of the wire and the lubricant film thickness after reaching steady state. It shows on the left side of Fig. 14 that the radial displacement starts to increase when the lubricant achieves the pressure peak. Moreover, the highest radial displacement at the FSI interface is situated just before the narrow gap between wire and die. When the wire leaves the die, the wire is exposed to a limited elastic recovery as expected (right side Fig. 14). Consequently, the minimum film thickness of 25 µm is situated at the outlet of the die. An important consequence of the evolution of the minimum film thickness during the FSI simulation is the calculation of the resulting drawing force. A force of *280 N* is obtained after reaching the steady state solution, which is a significant difference with the forces calculated in Table 3 and 4, where only CFD simulations are considered with a fixed minimum film thickness of 10 µm.

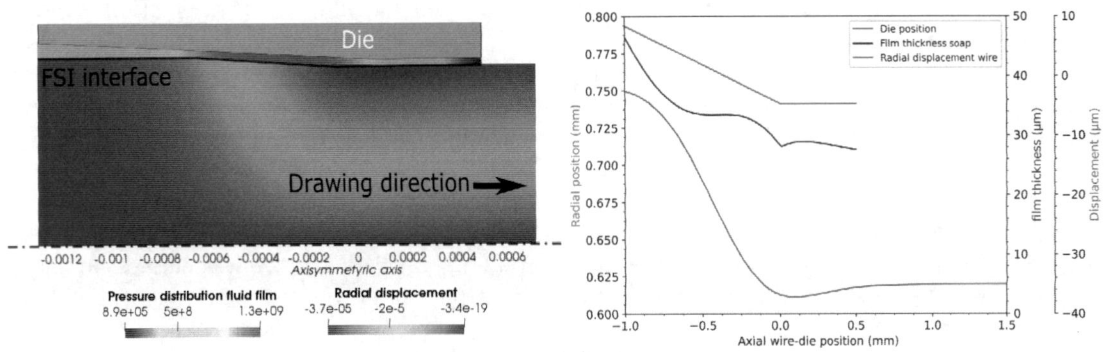

Figure 14: Left: radial displacement of the wire during deformation and pressure distribution lubricant; Right: Radial displacement wire at FSI interface and lubricant film thickness.

IOP Conf. Series: Materials Science and Engineering **1312** (2024) 012008 doi:10.1088/1757-899X/1312/1/012008

A viscous Newtonian flow at steady state between two walls can be described as a combination of pressure and viscosity driven flow. This viscosity and pressure driven flows are expressed as a Couette and Poiseuille flow [4, 5], respectively. Fig. 15 presents the behaviour of both flows between two infinitely long parallel parts in axial direction with a fixed distance h at steady state. The flows are examinated in the $x-r$-plane, as the velocity profile of the lubricant at a given axial position does not change in circumferential direction ($\frac{\delta \vec{u}}{\delta \phi}=0$) (Fig. 15). In a Couette flow one of the two walls has a velocity $\vec{u_{lp}}$ while the other one is at rest ($\vec{u}_{up}=0$). $\vec{u_{lp}}$ represents the velocity of the lower part and $\vec{u}_{up}$ is the velocity of the upper part. Assuming $r=r_{lp}$ is the radial position at the lower part and $r=r_{up}=r_{lp}+h$ the radial position at the upper part, the relative motion of the surfaces imposes a constant shear stress on the fluid over the radial direction and induces a flow in axial direction with a linearly varying velocity $\vec{u}=\vec{u_{lp}}(1-\frac{r}{r_{up}})$ with respect to the radial position r. In a Poiseuille flow both parts are at rest and the flow is forced by the pressure gradient in the direction parallel to the parts and presents a symmetric behaviour over the fluid in radial direction, which is demonstrated in Fig. 15.

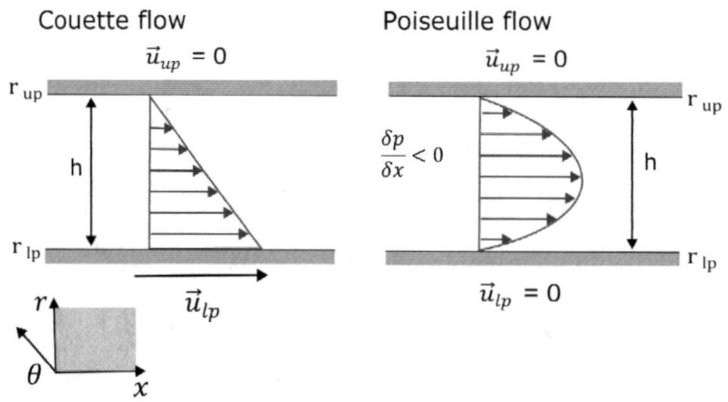

Figure 15: Couette and Poiseuille flow.

Fig. 16 illustrates the velocity profile over the radial direction of the lubricant at 5 positions across the fluid domain with the corresponding pressure and absolute value of the shear stress behaviour on the FSI interface. It can be noticed that the pressure increases significantly from the die inlet to position 3. This results in a positive pressure gradient and consequently increased backflow near the die boundary, which can be seen in the flow fields at position 0 and 1. Moreover, this influences partially the magnitude of the shear stress as the Couette flow becomes more significant on the FSI interface from position 1 to 2. Additionally, the slope of the shear stress magnitude becomes steeper around location x = -1.5 mm as the wire starts to deform, triggering an acceleration of the structure. Upon reaching the maximum pressure, the pressure gradient becomes zero at position 3, describing a pure Couette flow field. The further gradient after reaching the pressure peak describes a quasi-linear drop in pressure, which superimposes a Poiseuille behaviour on the Couette flow. Consequently, the shear stress decreases in magnitude, which can be observed in the flow field at position 4. This is followed by an increase in shear stress due to a strengthening of the Poiseuille effect, which is illustrated at position 5. When the flow arrives at the narrow gap under the die, a constant shear stress is observed until the outlet of the die is reached.

4.3. Analysis numerical performance of FSI simulation dry wire drawing

Finally, the coupling convergence is analysed. Fig. 17 displays the residuals of 6 time steps at the steady state condition. Each blue dot represents an FSI coupling iteration, with the

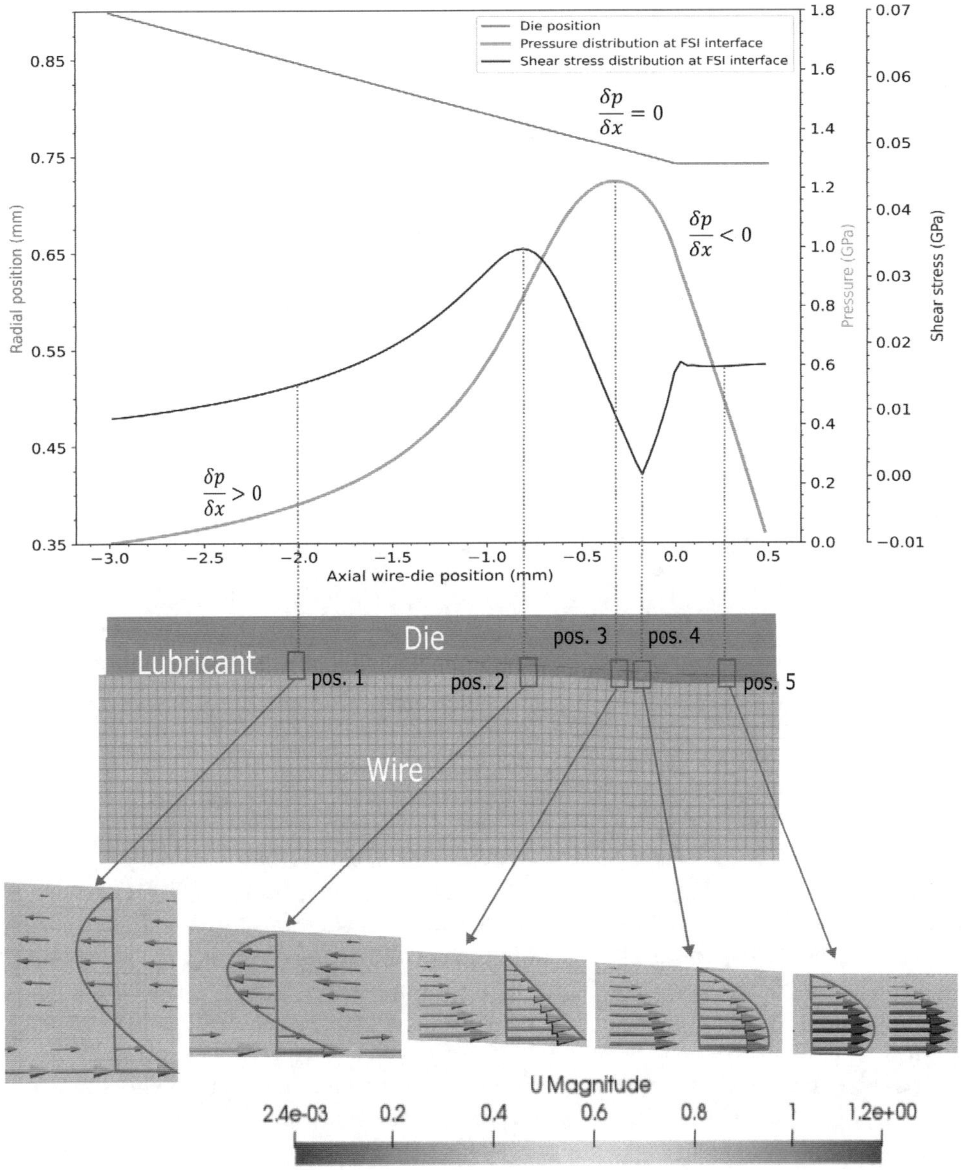

Figure 16: Pressure and absolute value of the shear stress distribution with corresponding velocity profile of the lubricant during wire drawing.

calculated residual. For each time step, the FSI tolerance of 0.005 µm has been reached. The average number of FSI iterations per time step at steady state is 3 with a reuse q = 0, which indicates a good performance of the IQN-ILS solver [18]. Moreover, 97.3% of the time is reserved for the structural calculation, while the fluid solver was 2.3% active and the FSI solver needs 0.4% calculation time. It has been observed that the computation time of the structural solver increases significantly with the increase in the number of wire cells reaching the yield criteria during the die diameter decrease. Further investigation is needed to improve the computational cost of the structural solver.

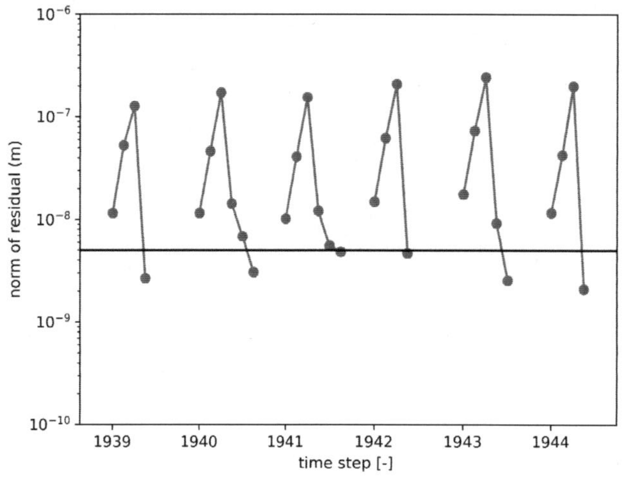

Figure 17: FSI Residuals at steady-state.

5. Conclusion

This work presents a 2D axisymmetric FSI simulation of the dry wire drawing procedure solved by the incompressible, isothermal pimpleFoam fluid solver in OpenFOAM 8 and a Lagrangian structural solver in foam-Extend 4.1. Both solvers are coupled by the quasi Newton technique IQN-ILS implemented in the in-house python based coupling tool CoCoNuT. A brief outline presents the strongly partitioned approach used in this coupling tool and reveals the communication between the OpenFoam solvers and CoCoNuT. Moreover, a detailed overview describes the approach of solving a complex problem like dry wire drawing numerically. This involves the successful implementation of the *timeVaryingMappedSolidtraction* and *timeVaryingMappedCoupledVelocity* boundary conditions in order to solve the sliding interface problem and applying the no-slip condition, respectively. Introducing the no-slip condition implies a better prediction of the energy consumption and the behaviour of the drawing parameters. Moreover, the introduction of the layering technique provides a reduction of the computational cost. However, the combination of layering and no-slip initially caused an oscillating behaviour in the tangential velocity component at the interface, which induced divergence of the FSI algorithm. These oscillations disappear by adding a layer of cells each time step at the inlet of the structural domain. Furthermore, the results show the behaviour of certain drawing parameters at steady state, with an estimated uncertainty based on grid refinement studies of 1.63%. Despite the Newtonian formulation of the fluid film, the flow field exhibits an expected behaviour consistent with the pressure and shear stress distribution of the lubricant. Moreover, it is qualitatively validated in terms of Poiseuille and Couette flow. Furthermore, the presented dry wire drawing simulation shows a converged solution with a good performance of the IQN-ILS solver. However, the structural solver has a significant larger time consumption with respect to the calculation time of the fluid and FSI solver, which requires further investigation.

Acknowledgments

The authors gratefully acknowledge financial support from NV Bekaert SA, Belgium through Dr. Stijn De Pauw.

References

[1] Montmitonnet P 2001 Plasto-hydrodynamic lubrication (PHD) - application of lubrication theory to metal forming process *C.R. Acad. Sci. Paris* **2** 729-37

[2] Reynolds O 1886 On the theory of lubrication and its applications to Mr. Beauchamps tower's experiments, including an experimental determination of the viscosity of olive oil *Philosophical transactions of the Royal Society of London* **177** 157-234

[3] Škurić V, De Jaeger P and Jasak H 2018 Lubricated elastoplastic contact model for metal forming processes in OpenFOAM *Comput. Fluids* **172** 226-40

[4] van Beek A 2019 *Advanced Engineering Design, Lifetime performance and reliability* (Delft: TU Delft) p 460

[5] Hamrock B J, Schmid S R and Jacobson B O 2004 *Fundamentals of fluid film lubrication, second edition* (New York: Marcel Dekker, Inc.) p 728

[6] Rajagopal K and Szeri A 2003 On an inconsistency in the derivation of the equations of elastohydrodynamic lubrication *Proceedings of the Royal Society of London. Series A: Mathematical, Physical and Engineering Sciences* **459** 2771-86

[7] Bair S, Khonsari M and Winer W O 1998 High-pressure rheology of lubricants and limitations of the Reynolds equation *Tribol. Int.* **31** 3323-41

[8] Hartinger M, Dumont M-L, Ioannides S, Gosman D and Spikes H 2008 CFD modeling of a thermal and shear-thinning elastohydrodynamic line contact *Journal of Tribology* **130** 041503-16

[9] Hajishafiee A, Kadiric A, Ioannides S and Dini D 2016 A coupled finite-volume CFD solver for two-dimensional elastohydrodynamic lubrication problems with particular application to rolling element bearings *j.tribol* **109** 258-73

[10] Wagner S, Munsch M and Delgado A 2022 An integrated OpenFOAM membrane fluid-structure interaction solver *OpenFOAM journal* **2** 48-61

[11] Cardiff P, Karač A, De Jaeger P, Jasak H, Nagy J, Ivanković A and Tuković Ž 2018 An open-source finite volume toolbox for solid mechanics and fluid-solid interaction simulations *Elsevier* **00** 1-45

[12] Tuković Ž, Karač A, Cardiff P, Jasak H and Ivanković A 2018 OpenFOAM finite volume solver for fluid-solid interaction *Transactions of FAMENA* **42** 1-31

[13] Havaej P, Degroote J and Fauconnier D 2023 A quantitative analysis of double-sided surface waviness on TEHL line contacts *j.triboint* **183** 1-19

[14] Havaej P, Degroote J and Fauconnier D 2023 Sensitivity of TEHL simulations to the use of different models for the constitutive behavior of lubricants *Lubricants* **11** 1-20

[15] St-Onge G and Olivier M 2023 Modular framework for the solution of boundary-coupled multiphysics problems *OpenFOAM journal* **3** 120-45

[16] Chourdakis G, Schneider D and Uekermann B 2023 OpenFOAM-preCICE: Coupling OpenFOAM with external solvers for multi-physics simulations *OpenFOAM Journal,* **3** 25

[17] Rodriguez S and Cardiff P 2022 A general approach for running Python codes in OpenFOAM using an embedded pybind11 Python interpreter *OpenFOAM journal* **2** 166-82

[18] Delaissé N, Demeester T, Fauconnier D and Degroote J 2021 Surrogate-based acceleration of quasi-Newton techniques for fluid-structure interaction simulations *Comput. Struct.* **260** 21

[19] Delaissé N, Demeester T, Haelterman R and Degroote J 2023 Quasi-Newton methods for partitioned simulation of fluid-structure interaction reviewed in generalized Broyden framework *Arch. Comput. Methods. Eng.* **30** 3271-3300

[20] Delaissé N 2023 *Development of fluid-structure interaction algorithms an application to elastohydrodynamic lubrication* (Ghent: Ghent University, Faculty of Engineering and Architecture) p248

[21] Dadvand P, Rossi R and Oñate E 2010 An Object-oriented Environment for Developing Finite Element Codes for Multi-disciplinary Applications *Arch Computat Methods Eng* **17** 253-97

[22] Vervaecke M, Fauconnier D and Degroote J 2022 OpenFOAM model of fluid-structure interaction in dry wire drawing, *ECCOMAS Congress, Oslo* **8** 11.

[23] Vervaecke M, Fauconnier D and Degroote J 2023 Fluid-structure interaction model of dry wire drawing with coupled axial velocity and layering *ECCOMAS Coupled Problems congres, Chania* **10** 12

[24] Holzmann T Mathematics, numerics, derivations and OpenFOAM, The basics for numerical simulations *Holzmann CFD* Release 7.0 URL https://Holzmann-cfd.de

[25] Cardiff P, Tuković Ž , De Jaeger P, Clancy M and Ivankovic A. 2017 A Lagrangian cell-centred finite volume method for metal forming simulation *Int. J. Numer. Meth. Engng,* **109** 1777-1803.

[26] Hwang J, Son I, Yoo J, Zargaran A and Kim N 2015 Effect of reduction of area on microstructure and mechanical properties of twinning-induced plasticity steel during wire drawing *Met. Mater. Int* **21** 815-22

[27] Breuer M, De Nayer G, Münsc M, Gallinger T and Wüchner R 2011 Fluid-structure interaction using a partitioned semi-implicit predictor-corrector coupling scheme for the application of large-eddy simulation *J. Fluids Struct.* **29** 107-30

[28] Greenshields C and Weller H 2022 *Notes on computational fluid dynamics: general principles* (UK: CFD Direct Limited) p291
[29] Versteeg H and Malalasekera W 2007 *An introduction to computational fluid dynamics, the finite volume method* (Harlow: Pearson) p503
[30] Hirsch C 2007 *Numerical computation of internal and external flows, volume 1 fundamentals of computational fluid dynamics* (Burlington: Butterworth-Heinemann) p656
[31] Maneeratana K 2000 Development of the finite volume method for non-linear structural applications *Ph.D. Thesis, Imperial College of Science, Technology and Medicine London*
[32] Cardiff P, Karač A and Ivanković A 2014 A large strain finite volume method for orthotropic bodies with general material orientations *Comput.Methods Appl.Mech.Engrg* **268** 318-35
[33] Karl T, Zartmann J, Dalpke S, Gatti D, Frohnapfel B and Böhlke T 2023 Influence of flow-fiber coupling during mold-filling on the stress field in short-fiber reinforced composites *Computational Mechanics, Springer* **71** 991-1013
[34] Jasak H and Tuković Ž 2010 Dynamic mesh handling in OpenFOAM applied to fluid-structure interaction simulations *ECCOMAS CFD congres, Lisbon* **5** 19
[35] ReFRESCO Accessed October 2023 *A community based open-usage and open-source CFD code for the Maritme World* (`http://www.refresco.org/verification-validation/utilitiesvv-tools/`)
[36] Eça L and Hoekstra M 2014 A procedure for the estimation of the numerical uncertainty of CFD calculations based on grid refinement studies *j.comput phys* **262** 104-30
[37] Mac Donald B J 2013 *Practical stress analysis with finite elements* (Dublin: Glasnevin)p388
[38] Simo J C and Hughes T J R 1997 Computational Inelasticity (Cambridge: Springer) p392

IOP Conf. Series: Materials Science and Engineering **1312** (2024) 012009 doi:10.1088/1757-899X/1312/1/012009

Understanding superlinear speedup in current HPC architectures

Flavio Cesar Cunha Galeazzo[1,*], **R. Gregor Weiß**[1], **Sergey Lesnik**[2], **Henrik Rusche**[2], **Andreas Ruopp**[1]

[1]High Performance Computing Center Stuttgart (HLRS), University of Stuttgart, Nobelstraße 19, D-70569 Stuttgart, Germany

[2]WIKKI GmbH, Ziegelbergsweg 68, D-38855, Wernigerode, Germany

E-mail: *flavio.galeazzo@hlrs.de

Abstract. The performance of OpenFOAM in strong scaling tests on HPC systems with AMD EPYC processors exhibits a pronounced superlinear speedup. Simple test cases show superlinear speedups of over 300%, which significantly impacts the efficient use of computing resources.

With the last generation of HPC architectures, a superlinear speedup of about 10% to 20% was well expected and accepted by CFD users [1]. The measured superlinear speedup is much more pronounced and withstands the communication overhead to even larger scales.

A detailed performance analysis of OpenFOAM follows, employing various High-Performance Computing (HPC) architectures, including AMD, ARM and Intel systems. The performance metric FVOPS (Finite VOlumes solved Per Second) is introduced to compare the performance of Computational Fluid Dynamics (CFD) applications when varying the grid size, as occurs in a strong scaling test. The achievable FVOPS depends on various factors, including the simulation type, boundary conditions, and especially the grid size of a use case. Analysing FVOPS on a single node level with varying grid size shows a significant difference in performance and cache utilization, which explains the large superlinear speedups seen in the strong scaling tests.

Furthermore, FVOPS can be used as a simple benchmark to determine the optimal number of grid elements per rank to simulate a given use case at peak efficiency on a given platform, resulting in time, energy, and cost savings.

The FVOPS metric also facilitates the direct comparison between different HPC architectures. The tests using AMD, ARM, and Intel processors show a peak in performance when employing around 10,000 grid elements per core. The presence of a large L3 cache on AMD processors is particularly advantageous, as indicated by L3 cache miss rates observed on AMD EPYC processors. Our results suggest that future HPC architectures with larger caches and higher memory bandwidth would benefit the CFD community.

1. Introduction

Computational Fluid Dynamics (CFD) accounts for a large share of the utilisation of High-Performance Computing (HPC) systems, particularly supercomputers. An example is the supercomputer Hawk from HLRS, where the proportion of projects using CFD is over 60% [2].

The exaFOAM project [3] aims at exploring the use of massively parallel HPC architectures for CFD simulations using the software OpenFOAM [4]. There is a particular interest in scalable simulations using different HPC architectures. The study focuses explicitly on AMD, ARM,

IOP Conf. Series: Materials Science and Engineering **1312** (2024) 012009 doi:10.1088/1757-899X/1312/1/012009

and Intel systems. The findings reveal a noteworthy phenomenon – specific HPC systems demonstrate a remarkable superlinear speedup, i.e., that the speedup is greater than the number of processors that are used [5].

The investigation encompasses two test cases: the exaFOAM Grand Challenge with the DLR Combustor and a more straightforward laminar flow case. A superlinear speedup ranging from 10% to 20% is usually seen when using HPC systems for CFD [1]; however, some current HPC architectures display much larger speedups. Notably, strong scaling tests on AMD EPYC processors exhibit superlinear speedups exceeding 300%. Such a large difference in performance impacts the efficient utilisation of computer resources.

The core methodology involves strong scaling tests in which the total number of grid elements remains constant while the number of computing nodes increases. This approach reduces the number of solved grid elements per rank as more compute nodes are added to the calculation. Recognizing that OpenFOAM and CFD applications are memory-bound by nature is essential. Hence, the bottleneck for CFD applications lies in the bandwidth with which the data is fed to the CPU in HPC architectures.

This study also investigates the reason for the observed superlinear scaling. Superlinear scaling is expected in scenarios where the available memory bandwidth increases with the number of nodes. However, we show that with the AMD EPYC platform, the better utilization of the L3 cache is the main reason for the superlinear speedup. It is also emphasised that the advantages resulting from the better use of resources come up against limits imposed by the communication overhead at larger scales. Using the AMD EPYC platform, the superlinear speedup withstands the anticipated communication overhead, extending its influence even to larger scales. This phenomenon is analyzed in detail, and its implications are explored.

2. Strong scaling tests

The strong scaling tests used the Hawk supercomputer from HLRS, which is an HPE Apollo system with 5632 compute nodes, each with dual socket AMD EPYC Rome 7742 processors (2 x 64 cores @ 2.25GHz), 256 GB of RAM and InfiniBand HDR200 network. The AMD EPYC CPUs have a hierarchical L3-cache configuration, with each socket having 16 Core Complexes (CCX), and each CCX having, in turn, four Zen2 cores and 16MB of shared L3-cache, making 256 MB per socket. OpenFOAM v2112 and v2212 were used in the benchmarks, compiled with GCC 10.2 and HPE MPT 2.26.

2.1. DLR Combustor

The considered combustor stems from an experimental setup constructed by Severin [6] at the Institute of Combustion Technology of the German Aerospace Center (abbreviated as DLR in German). The burner is based on the Recirculation-Stabilized Jet Flame (RSJF) concept, which is better known as FLOX. Initially developed for industrial furnaces, this technology has great potential in the application of gas turbines. Compared to the widespread swirl burners, RSJF combustors feature lower NOx emissions, more homogeneous temperature distribution, and operate with a broader range of fuels and loads. The main goal of the experimental work was to clarify how the recirculation contributes to flow stabilization. Apart from the experiments, several numerical investigations with the same geometry were performed, e.g. by Ax et al. [7] and Gruhlke et al. [9]. The exaFOAM Grand Challenge GC2 test case files [12] and grid and restart files [13] are available.

We consider here the strong scaling with the case consisting of 489 million cells. The metric used is wall-clock time per time step, whereby I/O and initialization (i.e. the first time step) were not accounted for. The smallest run was performed on 8 nodes (1,024 cores) due to memory limitations, and the largest run with 256 nodes (32,768 cores). Figure 2 shows two series of measurements. The "pre-exaFOAM" series demonstrates the speed-up that was measured

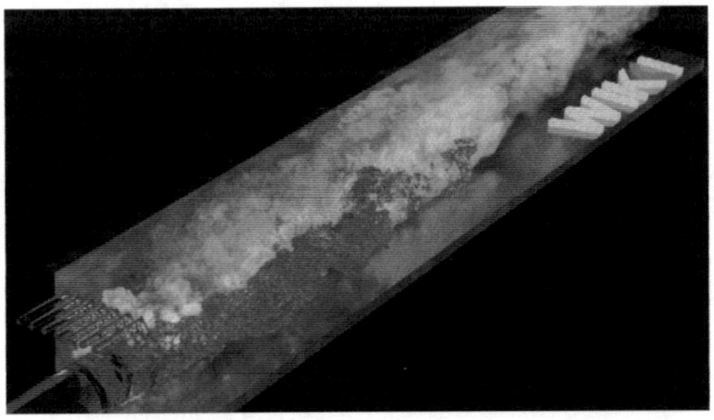

Figure 1. Rendering of the DLRCJH combustor flame with fuel.

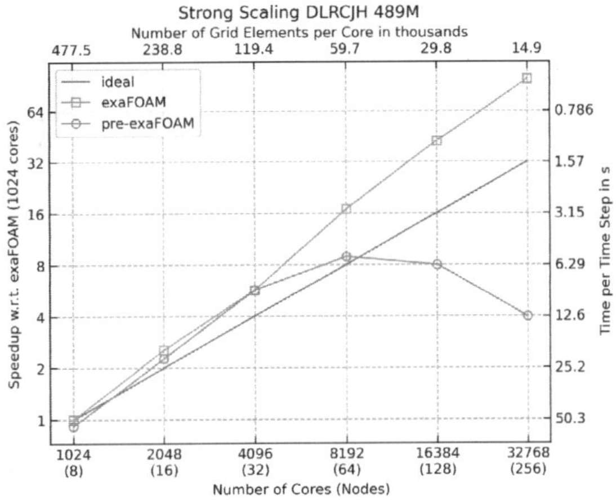

Figure 2. Strong scaling test done with the Hawk supercomputer for the 489M grid.

before the improvements made by the exaFOAM project. Due to this measurement, the most efficient run was on 32 nodes with a superlinear speedup of 42% over the ideal one. The most performant run was on 64 nodes. Profiling of the case identified an inefficiency in the algorithm responsible for the turbulence modelling, which triggered expensive MPI communication every time step. An algorithmic improvement drastically reduced the amount of MPI communication. The corresponding speedup is presented under the label "exaFOAM". With the improvements, the speedup is superlinear for all decompositions using more than 8 nodes. The most efficient is at the same time the run with the highest performance carried out on 256 nodes with the superlinear speedup of 98 being 206% higher than the corresponding ideal speedup of 32. Comparing performance before and after the improvement, the total time-to-solution is decreased by one order of magnitude, whereby the highest speedup improved from 10 (64 nodes) to 98 (256 nodes).

2.2. Lid-driven cavity

The flow inside a cavity driven by the movement of its upper boundary (lid) is one of the most standardized and ubiquitous test cases for CFD software (see Figure 3). Its simple geometry and boundary conditions make the test case easy to set up and validate. The lid-driven cavity has been used for more than 50 years to test and benchmark fluid dynamics codes [10], and has been adapted in the exaFOAM microbenchmark MB1. The test case files are available at

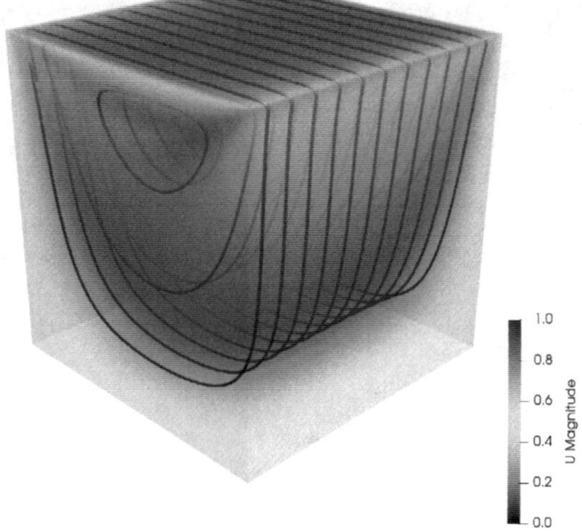

Figure 3. Visualization of the 3D lid-driven cavity flow.

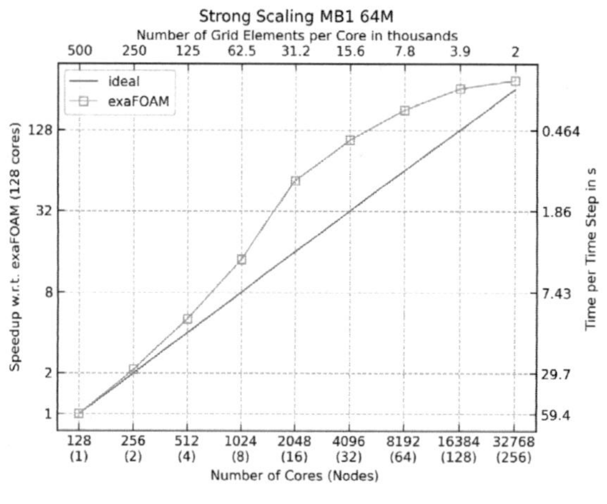

Figure 4. Strong scaling benchmark for the lid-drive cavity case with pronounced superlinear speedups.

Ref [11].

The strong scaling study employed the test case variant with a fixed number of iterations and 64 million grid elements. Figure 4 shows the strong scaling results, with the speedup with 32 nodes (4,096 cores) being 340% of the ideal (linear) speedup. The results maintain a superlinearity for up to 256 compute nodes (32,768 cores).

3. Single node performance

The performance of OpenFOAM in the AMD EPYC architecture with increasing grid size is analyzed to support the findings of the strong scaling. For this purpose, the performance metric FVOPS (Finite VOlumes solved Per Second) is introduced, similar to the MLUPS (Mega Lattice

Total grid elements	Runtime per time step [s]	Grid elements per rank	FVOPS per node	L3 ratio	MPI ratio	Parallel efficiency	IPC	IPC*PE
27,000,000	8.651	210,938	3.1M	0.823	0.056	0.96	0.31	0.299
15,625,000	4.267	122,070	3.6M	0.703	0.067	0.95	0.4	0.379
8,000,000	1.599	62,500	5.0M	0.520	0.089	0.92	0.73	0.643
5,359,375	0.646	41,870	8.3M	0.284	0.187	0.82	1.41	1.096
3,375,000	0.261	26,367	12.9M	0.083	0.279	0.77	1.72	1.238
1,953,125	0.144	15,259	13.6M	0.054	0.353	0.71	1.95	1.214
1,000,000	0.075	7,813	13.4M	0.050	0.469	0.65	2.01	1.088
421,875	0.040	3,296	10.5M	0.095	0.625	0.54	2.07	0.876
125,000	0.025	977	5.0M	0.266	0.792	0.37	1.82	0.536
15,625	0.017	122	0.9M	0.342	0.875	0.19	1.33	0.324

Table 1. Results for grids sizes from 27M to 15k elements. The different metrics are explained in the following sections.

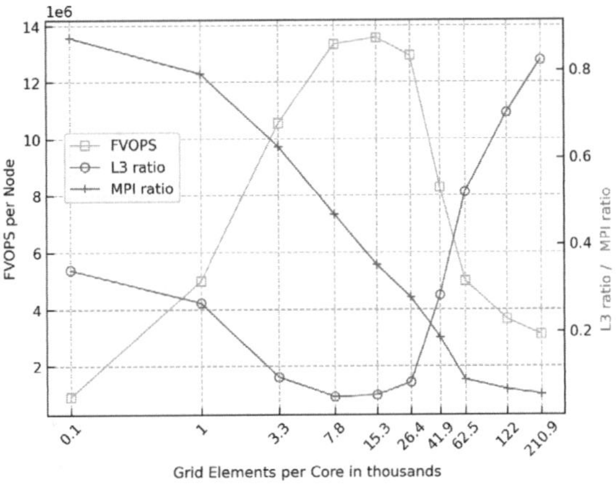

Figure 5. FVOPS, L3 ratio and MPI ratio for grids sizes from 27M to 15k elements, using one node.

Updates Per Second) metric used for Lattice Boltzmann codes [19]. The FVOPS metric is defined as:

$$\text{FVOPS} = \frac{\text{number of finite volumes of the grid}}{\text{runtime per time step or iteration in seconds}}.$$

For unsteady simulations, the runtime per time step is used, while for steady-state simulations, the runtime per iteration is the chosen divisor. By combining simulation size and runtime, FVOPS offers a comprehensive performance evaluation. As the MLUPS metric, FVOPS reflects the system's efficiency in CFD applications. Various simulation parameters affect the FVOPS metric, including grid size, partitioning, parallel efficiency, and the number of variables being solved. It is also a versatile metric for cross-system comparisons when using the same set of parameters.

Focusing on the FVOPS results in Figure 5 and Table 1, the difference in performance between the simulation with the largest grid (27M) and the sweet spot (1.9M) with approximately 15k grid elements per rank is 438%. This result supports the superlinear speedup seen in the strong

	CPUs	Cores per node	Memory	L3 cache
Hawk	AMD EPYC 7742 @ 2.25 GHz	2 x 64	256 GB	256 MB
Vulcan	Intel Xeon Gold 6138 @ 2,0 GHz	2 x 20	192 GB	27.5 MB
Armida ARM	Cavium ThunderX2 @ 2.5 GHz	2 x 32	256 GB	32 MB
Armida AMD	AMD EPYC 7313 @ 3.0 Ghz	2 x 16	256 GB	128 MB
Armida Intel 1	Intel Xeon Gold 6330 @ 2.0 GHz	2 x 28	256 GB	42 MB
Armida Intel 2	Intel Xeon Gold 6326 @ 2.9 GHz	2 x 16	512 GB	24 MB

Table 2. Hardware of the computer systems used in the performance analysis.

scaling test, as the system performance increases dramatically with the decrease in grid size.

In the strong scaling test with the lid-drive cavity case (Figure 4), the superlinearity is highest with 15k elements per core, which correlates well with the FVOPS performance in a single node. Although being a different setup, the DLR combustor results (Figure 2) show the same trend, with clearly superlinear scalability with 29k or 15k grid elements per core.

4. Analysis using different architectures

The analysis using the AMD EPYC 7742 showed a pronounced difference in performance when using grids of different sizes. To evaluate the impact of the CPU architecture on the performance, different architectures were studied, described in Table 2. Although the same case configuration and sizes have been used in all platforms, the amount of grid elements per core varies, as each system has a different core count per node.

In Figure 6, FVOPS is calculated per node, as this is the most important metric. However, the amount of grid elements per core is used in the x-axis to facilitate the comparison. Most systems show peak performance using around 10k grid elements per rank, with only the Armida AMD system showing better performance at around 50k grid elements per rank. In the systems using Intel, ARM, and the Armida AMD CPU, the performance difference is between 140% and 250% when using the largest grid size and the grid with peak performance. This difference is more pronounced using the Hawk system with AMD EPYC 7742 CPU, with the peak performance being 438% higher. It can be argued that the more significant difference in the performance with the AMD EPYC 7742 CPU is due to its larger core count (64 cores per socket) and its larger L3 cache (256 MB).

In summary, the AMD EPYC 7742 CPU platform has a significant performance advantage. The performance of the other platforms is lower, especially when using less than 60k grid elements per core.

5. Profiling of L3 cache and MPI overhead

It is often claimed that the reason for the superlinear speedup is that more memory bandwidth is available in parallel execution [1, 5]. In the particular example being analyzed, however, it is argued that the better utilization especially of the L3 cache is the main reason for the superlinear speedup. To study the behaviour of the system regarding the L3 cache utilization, the L3 ratio is defined as the ratio between L3 cache misses and hits:

$$\text{L3 ratio} = \frac{\text{L3 cache misses}}{\text{L3 cache hits}}.$$

The measurements using L3 cache counters with the AMD EPYC 7742 are, unfortunately, not as accurate as they could be. The L3 ratio has been measured using the tool perf [14], accounting for total (system and user) L3 cache hits and misses. The AMD Zen2 platform does

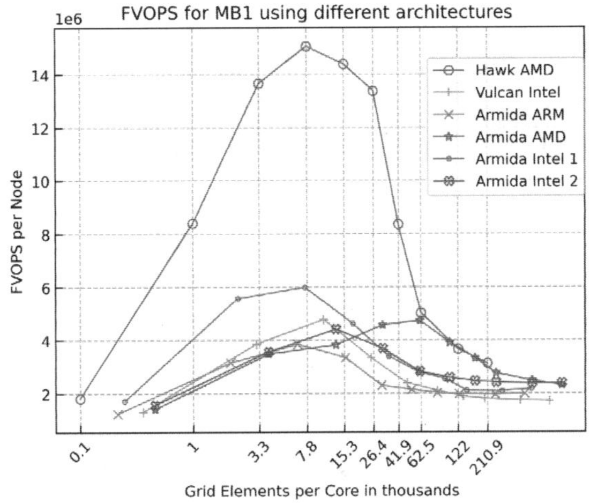

Figure 6. FVOPS using different architectures.

not differentiate between L3 cache counters from the system and user spaces. That prevents the use of PAPI counters [15], for example, as PAPI uses counters from the user space. This can be different in new AMD architectures. Nevertheless, user space L3 cache hits and misses are likely to dominate the total cache metric, as the compute nodes in Hawk use an operating system with minimal system processes.

Although not as accurate as L3 cache counters from the user space, the L3 ratio using total counters is shown in Figure 5 and Table 1. With the largest grid, the L3 ratio is over 80%, showing that, at this point, the performance is limited by the main memory access. On the contrary, at the sweet spot, the L3 ratio is minimal, which translates into less access to the main memory and, consequently, more useful computation by the compute cores. These results corroborate the finding that the system performance increases with a decrease in grid size, but even considering that the L3 ratio increases at grid sizes below the sweet spot, it still does not explain the pronounced drop in performance at the smallest grid sizes.

To investigate the drop in performance at the smallest grid sizes, which imply small memory footprints and short runtimes, the MPI communication has been profiled using the tool mpiP [16]. Figure 5 shows the MPI ratio as the fraction of compute time and time spent in MPI communication. The MPI ratio increases as the grid size decreases. It is observed that as the runtime per time step gets shorter (for 15k grid elements it is about 0.007s), the importance of the MPI latency increases, being finally dominant.

6. Modeling the FVOPS behavior using POP metrics

To measure L3 cache counters, the user has to have administrator privileges in most systems. In the case the user has no access to the L3 cache counters, the use of POP metrics [17] measured with the tool extrae [18] is an option to evaluate the performance. Figure 7 shows the FVOPS for one node, along with two of the POP metrics: Instructions per cycle (IPC) and parallel efficiency (PE). IPC represents the number of instructions executed by the CPU for each clock cycle. Lower values of IPC indicate a low rate of computation, which typically comes from stalled processes waiting for data. PE measures the efficiency of the distributed computation and the communication associated with it. PE is a compound metric defined as the product of the load balance and the communication efficiency.

The IPC follows the pattern of FVOPS when decreasing the grid size, although the IPC does

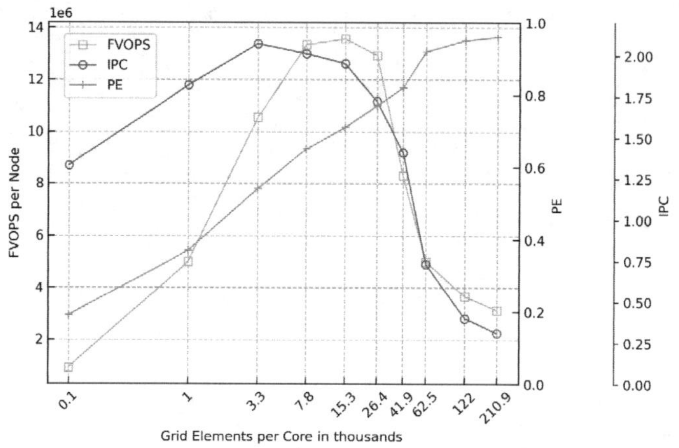

Figure 7. FVOPS, IPC, and PE for grids sizes from 27M to 15k elements, using one node.

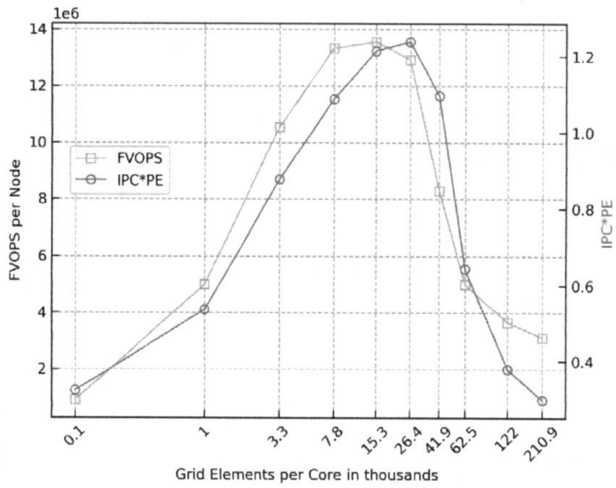

Figure 8. FVOPS and IPC*PE for grids sizes from 27M to 15k elements, using one node.

not drop significantly for small grid sizes as the FVOPS. The PE, on the other hand, decreases steadily with the decrease of the grid size. When multiplying these two metrics, the new metric IPC*PE shows a similar behaviour as measured with FVOPS (see Figure 8), and can also be used as a guide to finding the sweet spot in a particular system. The POP metrics can also be seen in Table 1.

7. Summary

Strong scaling tests showed significant superlinear speedups of up to 340% using OpenFOAM on the Hawk supercomputer, which is much higher than that obtained with the previous generation of computing platforms.

The driving force behind the observed superlinear speedups is the better performance of simulations with a specific grid size. This outlines that all platforms, regardless of processor type, exhibit an optimal performance threshold at a specific grid element per rank ratio. This knowledge can guide CFD practitioners seeking to optimize the performance of their simulations on diverse HPC architectures.

A contribution of this study is the proposal of the FVOPS metric, which allows the direct comparison of the system performance when changing the simulation grid size. FVOPS also serves as a practical tool to determine the optimal number of grid elements per rank on a specific HPC architecture. With this metric, users can make informed decisions about the optimal number of compute nodes to use for a given simulation, which translates into tangible savings in time, energy, and cost, making it a valuable tool.

The Hawk supercomputer uses AMD EPYC Rome 7742 CPUs, which showed a significant performance advantage over the other platforms in this study, including Intel, ARM and other AMD platforms. The large performance difference observed between runs with very large grid sizes and with the optimal number of grid elements per rank explains the significant superlinear speedup seen in the strong scaling tests.

Finally, the results indicate that future HPC architectures with larger L3 cache sizes and higher memory bandwidth would benefit the CFD community.

7.1. Acknowledgments

The simulations were performed on the national supercomputer HPE Apollo Hawk at the High Performance Computing Center Stuttgart (HLRS) under the grant number HLRSexaFOAM/44230. This work is carried out in the scope of the exaFOAM project, which has received funding from the German Federal Ministry of Education and Research and the European High-Performance Computing Joint Undertaking (JU) under grant agreements No 16HPC024 and No 956416, respectively. The JU receives support from the European Union's Horizon 2020 research and innovation programme and France, Germany, Spain, Italy, Croatia, Greece, and Portugal.

References

[1] Axtmann, G., Rist, U. (2016). Scalability of OpenFOAM with Large Eddy Simulations and DNS on High-Performance Systems. In: Nagel, W.E., Kröner, D.H., Resch, M.M. (eds) High Performance Computing in Science and Engineering ´16. Springer, Cham. `https://doi.org/10.1007/978-3-319-47066-5_28`

[2] C. M. Williams, E. Gedenk. HLRS Jahresbericht 2022, 2023, `https://www.hlrs.de/de/about/profile/jahresbericht`, accessed on 31.01.2024

[3] exaFOAM project website `https://exafoam.eu`, accessed on 30.01.2024

[4] H. G. Weller, G. Tabor, H. Jasak, C. Fureby. A tensorial approach to computational continuum mechanics using object-oriented techniques. Comput. Phys. 12 (6): 620–631, 1998. `https://doi.org/10.1063/1.168744`

[5] S. Ristov, R. Prodan, M. Gusev and K. Skala. Superlinear speedup in HPC systems: Why and when?, 2016 Federated Conference on Computer Science and Information Systems (FedCSIS), Gdansk, Poland, pp. 889-898, 2016.

[6] Severin, M., 2019. Analyse der Flammenstabilisierung intensiv mischender Jetflammen für Gasturbinenbrennkammern (PhD Thesis). Universität Stuttgart.

[7] Ax, H., Lammel, O., Lückerath, R., Severin, M., 2020. High-Momentum Jet Flames at Elevated Pressure, C: Statistical Distribution of Thermochemical States Obtained From Laser-Raman Measurements. Journal of Engineering for Gas Turbines and Power 142, 071011. `https://doi.org/10.1115/1.4045483`

[8] Gruhlke, P., Janbazi, H., Wlokas, I., Beck, C., Kempf, A.M., 2020. Investigation of a High Karlovitz, High Pressure Premixed Jet Flame with Heat Losses by LES. Combustion Science and Technology 192, 2138–2170. `https://doi.org/10.1080/00102202.2020.1781101`

[9] Gruhlke, P., Janbazi, H., Wollny, P., Wlokas, I., Beck, C., Janus, B., Kempf, A.M., 2021. Large-Eddy Simulation of a Lifted High-Pressure Jet-Flame with Direct Chemistry. Combustion Science and Technology 1–25. `https://doi.org/10.1080/00102202.2021.1903886`

[10] AbdelMigid, T. A., Saqr, K. M., Kotb, M. A., and Aboelfarag, A. A. Revisiting the lid-driven cavity flow problem: Review and new steady state benchmarking results using GPU accelerated code. Alexandria

engineering journal, 56(1), 123-13, 2017.

[11] MB1 Microbenchmark - Lid-driven cavity 3D `https://develop.openfoam.com/committees/hpc/-/tree/develop/incompressible/icoFoam/cavity3D`, accessed on 30.01.2024

[12] GC2 Grand Challenge - DLR Confined Jet High Pressure Combustor `https://develop.openfoam.com/committees/hpc/-/tree/develop/combustion/XiFoam/DLRCJH`, accessed on 30.01.2024

[13] Lesnik, Sergey and Rusche, Henrik, exaFOAM Grand Challenge GC2 - DLR Confined Jet High Pressure Combustor, `https://doi.org/10.18419/darus-3699`, DaRUS, V1

[14] `https://perf.wiki.kernel.org`, accessed on 30.01.2024

[15] `https://icl.utk.edu/papi`, accessed on 30.01.2024

[16] `https://github.com/LLNL/mpiP`, accessed on 30.01.2024

[17] `https://pop-coe.eu/node/69`, accessed on 30.01.2024

[18] `https://tools.bsc.es/extrae`, accessed on 30.01.2024

[19] Jingsheng Ma, Kejian Wu, Zeyun Jiang, and Gary D. Couples. SHIFT: An implementation for lattice Boltzmann simulation in low-porosity porous media, Phys. Rev. E 81, 056702, 2010.

Vertical axis turbine simulations based on sliding and overset meshes

Asmelash Haftu Amaha[1], Shivasubramanian Gopalakrishnan[1], Prabhu Ramachandran[1] and Joel Guerrero[2]

[1]Indian Institute of Technology Bombay, Powai 400076, Maharashtra, India
[2]Wolf Dynamics, Genoa 16127, Liguria, Italy

E-mail: asmelash.a@aero.iitb.ac.in

Abstract. Vertical Axis Turbines (VATs) are gaining traction as decentralized energy sources due to their simple design and suitability for various wind conditions. However, understanding VAT aerodynamics requires robust computational methods. This study compares the accuracy and efficiency of two meshing techniques (sliding mesh and overset mesh) for VAT simulations using OpenFOAM. Validation against experimental data confirms the capability of both methods to capture flow physics and predict turbine performance. The results suggest both techniques offer efficient and accurate VAT simulations, providing valuable tools for future VAT optimization and experimental validation.

1. Introduction

The horizontal axis wind turbine (HAWT) and vertical axis wind turbine (VAWT or VAT) are the leading devices for harnessing renewable energy from the wind. Currently, HAWTs are more advanced with improvements in size and aerodynamic efficiency, achieving a power coefficient close to the 50% range which is about 85% of the Betz limit [1, 2]. They are, however, very expensive to achieve affordable and distributed modern energy systems and sustainability goals [3]. VATs also called cross-flow turbines were widely studied by Sandia from 1970 to 1990, and a resurgence of interest is seen from 2010 till date. VATs can have one or more straight or curved blades that rotate parallel to the axis of rotation.VATs have some attractive features compared to HAWTs. They are compact, independent of the flow direction, the mass is concentrated near ground level hence low operation and maintenance (OM) costs, and used for a wide range of applications including adaptability for small-scale domestic installations. In addition, VATs can supplement HAWTs or be employed in places where HAWTs are less effective, for example in Class 2 wind sites. Class 2 wind site can be considered as a moderately breezy location (5.4 to 6.4 m/s) that is not quite as windy as Class 3 or above. This

IOP Conf. Series: Materials Science and Engineering **1312** (2024) 012010 doi:10.1088/1757-899X/1312/1/012010

may not be the windiest site around, but they still offer the potential for wind energy development and are major wind sites in many countries of the world [4].

VAT blades experience complex unsteady aerodynamics, largely due to the blade motion inducing pressure differences in the vicinity. These pressure fluctuations interact with the freestream velocity in the upstream region and with the wake in the downstream region of the rotor. This unsteadiness is inherent and persists even if the wind speed is constant or the rotor speed is constant. Due to such complexities, accurate simulation of VAT performance is a challenging task and requires accurate meshing techniques [5, 6, 7, 8] and [9, 10, 11].

Numerous analytical and computational models have been developed in the past to analyze the aerodynamic performance of several VAT designs, ranging from the simple Blade Element Momentum (BEM) model approaches to more advanced CFD simulations. While CFD models are essential for accurate predictions, their reliability relies heavily on factors such as mesh generation technique, turbulence model selection, and appropriate boundary conditions. Among the existing literature, the sliding mesh technique is predominantly utilized for CFD simulations of VATs [2, 12]. However, it is noteworthy that only a limited number of studies have explored alternative techniques to the sliding mesh approach such as the overset mesh [13, 14] and morphing mesh [13, 15].

OpenFOAM is a powerful and flexible open-source computational fluid dynamics (CFD) code platform for simulating fluid flow and related phenomena in computational continuum mechanics [16]. Since its initial release in 2004, OpenFOAM is now a popular open-source CFD platform with a strong community of users and developers, accounting for 39% of all the CFD software usage till 2018 in wind energy conversion systems [17].

The dynamic mesh capability of OpenFOAM enables the simulation of moving objects and changing boundaries, making it a valuable tool for a wide range of CFD simulations involving complex motion scenarios. It provides several dynamic mesh classes such as adaptive mesh refinement (AMR), moving meshes (Whole domain), morphing meshes (Smoothing), sliding meshes (SM), and overset meshes (OM). In this work, we employed both SM and OM techniques for simulating vertical axis turbines.

The motivation and objectives of this study are (1) exploring the complex aerodynamics of vertical axis turbines, and (2) using computational fluid dynamics to analyze these sophisticated systems. It aims to explore vertical-axis turbines and advance computational modeling methods in this area by delving into the intersection of aerodynamics and CFD. In CFD, the sliding mesh technique is an established simulation model for many complex systems. Due to its flexibility to use for more complex geometries, the overset mesh technique is gaining more attention in academia and industries. This inspires us to explore its capability for accurately simulating VATs, which are also the research targets for decentralized and sustainable energy supply. The main focus is to provide a thorough comparison between sliding mesh and overset mesh methods by underlining the differences, advantages, and limitations of each. The comparison focuses on convergence and sensitivity studies, investigation of the flow field parameters, and the analysis of aerodynamic and performance parameters, which can contribute to understanding complex flow phenomena and optimizing simulation approaches.

2. Blade Aerodynamic Forces

The problem chosen for this comparison is the VAT based on Darrieus type H-rotor commonly employed for wind and hydrokinetic applications on small scales. Figure 1 shows the schematic diagram of three commonly used three bladed vertical axis turbines (VATs) used in wind energy. These are (a) H-Darrieus turbine, (b) Darrieus turbine, and (c) Helical-Darrieus turbine. The vertical axis supports the rotor with blades spinning perpendicular to the ground, like a cylinder with wings. The wind pushes or pulls on the blades, causing them to rotate around the vertical axis.

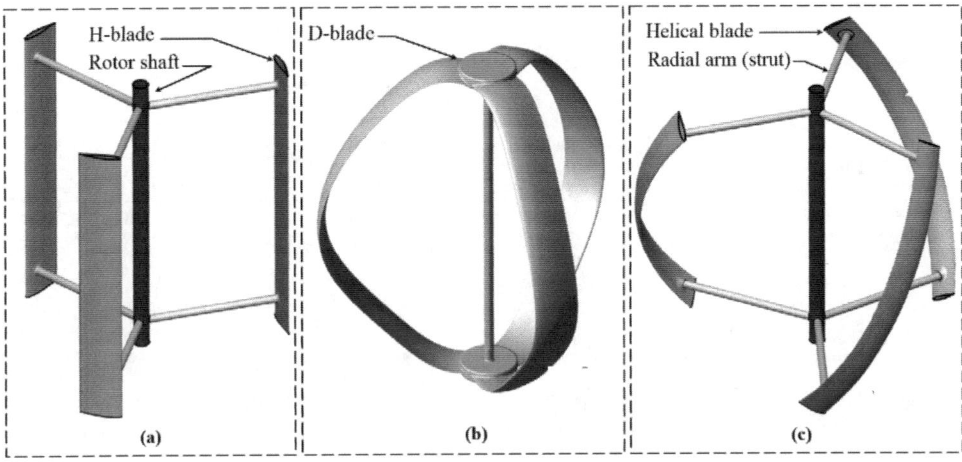

Figure 1: Schematic diagram of vertical axis turbines (VATs) used in wind energy. (a) H-Darrieus turbine, (b) Darrieus turbine, and (c) Helical-Darrieus turbine [18].

The aerodynamic efficiency of a vertical axis wind turbine (VAT) is expressed by the variation of the power coefficient curve, C_p, with a dimensionless tip speed ratio (TSR or λ). It is the ratio of extracted power to the available power. TSR and solidity (σ) are other commonly used dimensionless parameters to describe the performance and operation of a VAT blade are

$$C_p = \frac{P}{\frac{1}{2}\rho U_0^3 A_s}, \quad \lambda = \frac{\omega R}{U_0}, \quad \sigma = \frac{Nc}{R}, \tag{1}$$

where ρ is the density, U_0 is the free stream velocity, $A_s = DH = 2RH$ is area swept by the rotating blade, R is radius of the turbine ($D/2$), H is the height, P is power extracted by the turbine, ω is the angular velocity in rad/s, c is the chord length, and N the number of blades.

The Reynolds number (Re) is also obtained using the chord length and the blade tip speed. It can be defined as

$$Re = \frac{U_t c}{\nu}, \quad U_t = \omega R = U_0 \lambda, \tag{2}$$

where ν is the kinematic viscosity of the fluid and U_t is blade tip speed.

Figure 2 illustrates the schematic diagram of the top view of a single turbine rotor along its flight path, which is H-type Darrieus VAT with symmetric NACA blades. The upwind and downwind regions as well as the different flow segments are discussed in a later section. Figure 3 shows the aerodynamic forces acting on the VAT rotating in the counterclockwise direction. For, a multi-blade turbine, the rotor blades, each of height H, are initially evenly spaced at a radial distance R around the central axis, which is the vertical shaft. Throughout the rotation of the turbine (see Figure 3), airflow dynamics

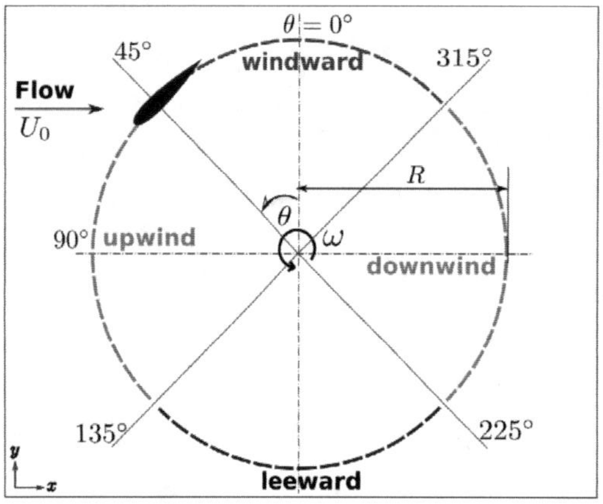

Figure 2: Schematic diagram for a VAT simulation and the computational domain with fixed and static components and boundary conditions.

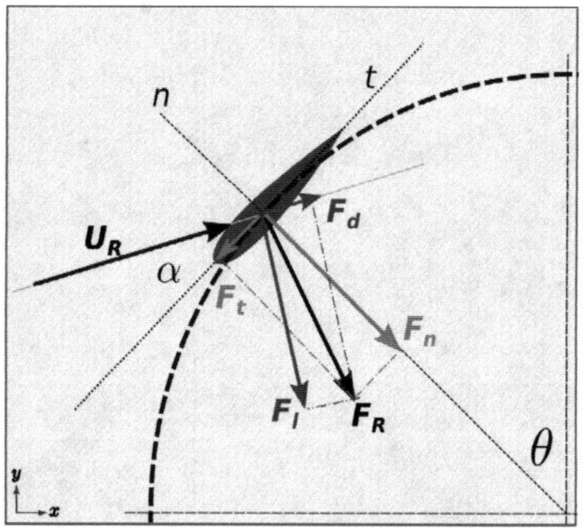

Figure 3: Schematic diagram and aerodynamic forces at a specific azimuth angle of VAT. Blade motion starts from $\theta = 0°$.

around the blades generate reaction forces, such as the lift F_l, drag F_d, normal force F_n and tangential forces F_t. The force, F_R is the resultant of the forces and U_r is the relative velocity. The dimensionless normal and tangential coefficients of these forces are revised as shown,

$$\begin{cases} C_n &= F_n/(0.5\rho U_0^2 A_b), \\ C_t &= F_t/(0.5\rho U_0^2 A_b), \end{cases} \tag{3}$$

where F_n and F_t are the normal and tangential forces of the blade flight path, and C_n and C_t are the coefficients, and F_l and F_d are obtained as a function the relative velocity and angle of attack.

3. Computational method

3.1. Governing Equations

Cells of small volumes form a mesh to describe the solution domain and in CFD simulations and in many cases the mesh remains stationary. In the case of a dynamic mesh, the mesh changes based on the motion of the solution domain or it changes topologically, this feature also works when a domain is decomposed for parallel simulation.

Utilizing the finite volume method (FVM), OpenFOAM solves the Reynolds transport theorem (RTT) in the integral form (as depicted in equation 4) [19, 20, 21, 22]. It represents the general transport equation for the flow property ϕ.

$$\int_{V_p} \frac{\partial \rho \phi}{\partial t} dV + \int_{V_p} \nabla \cdot [\rho \phi (\mathbf{u} - \mathbf{u}_g)] dV - \int_{V_p} \nabla \cdot (\rho \Gamma_\phi \nabla \phi) dV = \int_{V_p} S_\phi \phi dV, \tag{4}$$

where $\mathbf{u}$, p, and ρ are the flow velocity, the fluid pressure, and density, respectively. $\mathbf{u}_g$ is the mesh velocity in dynamic meshes which modifies the convective part of the equation, and the volume dV is a function of time indicating that shape of the cells is changing in time.

By considering the assumptions for incompressible unsteady Navier-Stokes equations (NSE) and accounting for a time-varying computational domain with a moving mesh, equation 4 can be cast into the continuity and momentum equations, as expressed in equations 5 and 6.

$$\nabla \cdot \mathbf{u} = 0 \tag{5}$$

$$\frac{\partial \mathbf{u}}{\partial t} + [(\mathbf{u} - \mathbf{u}_g) \cdot \nabla]\mathbf{u} = -\nabla p/\rho + \nu \nabla^2 \mathbf{u} \tag{6}$$

The sliding mesh and overset mesh techniques are methods for handling cases with dynamic mesh [19, 23]. The two distinct dynamic mesh techniques can be used to simulate VATs and both are briefly introduced in this section. They are convenient and able to simulate the movement of the VAT blades traversing a circular trajectory. The sliding mesh is first applied for the simulation (and validation) of the baseline VAT case. Both methods are discussed along with the CFD model of the turbine and configuration parameters shown in Table 1.

3.2. VAT Configuration and CFD Model

The Darrieus vertical axis turbine (VAT) is numerically investigated in 2D employing the experimental conditions of reference [24]. The simulation involves a single blade with a NACA0015 airfoil, 1.22 m diameter, and a rotational speed of 0.75 rad/sec. These parameters correspond to a Reynolds number of 67,000 and a tip speed ratio of 2.5, with an undisturbed water velocity of 0.183 m/sec. The rotor solidity, defined as the ratio of the airfoil chord to rotor radius, is 0.25, with a chord length of 0.1524 m. Vertical axis turbine with only one blade is not common in the industry, however, it is justifiable to use for research because it makes it more convenient to analyze the flow past its rotor, the aerodynamic blade loads, and the wake characteristics. The inclusion of additional blades is straightforward afterwards. In addition, reference data for validation is available in reference [24]. The parameters of the CFD model and those of the Oler experiment are summarized in Table 1.

3.3. Computational Domain and Boundary Conditions

The 2D computational domain represents a virtual wind tunnel with fixed and rotating components. An AMI interface facilitates mesh motion for the Sliding Mesh (SM) case (Figure 4). In contrast, the Overset Mesh (OM) case (Figure 5) only rotates a portion of the airfoil, eliminating the need for a full rotating core. Both methods utilize blades

Table 1: Parameters for VAT CFD Model

Parameters	Value	Parameters	Value
Airfoil	NACA0015	Reynolds number	67,000
Number of blade, N	1	Tip speed ratio, λ	2.5, 5.1, 7.5
Chord length, c	0.1524 m	Freestream velocity	0.183 m/s
Diameter turbine, D_t	1.22 m	Rotational speed, ω	0.75 rad/s
Diameter of Rotor, D_c	1.5 m	Blade Aspect Ratio	6.5617

(NACA profile) created using python, and the rotor STL (Stereolithography) is generated using Onshape ([25]). The blades act as obstacles within the domain and can perform various motions depending on the case, including rotating for Vertical Axis Turbines (VAT) or flapping for airfoils. The rectangular domain has defined inlet and outlet boundaries on the left and right faces, respectively, while the top and bottom represent the virtual wind tunnel walls.

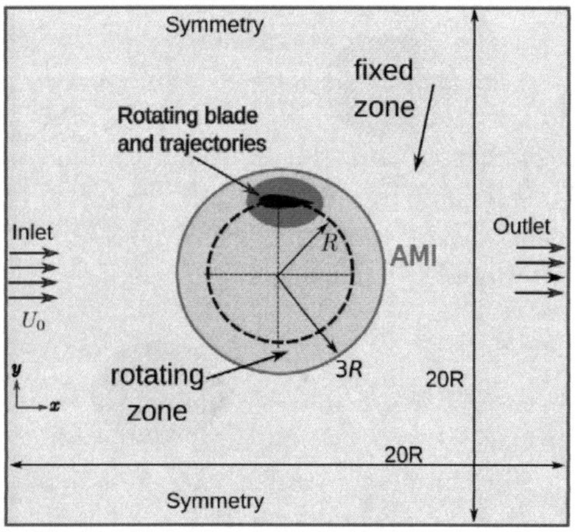

Figure 4: Schematic diagram, computational domain, and boundary conditions of a VAT simulation. Fixed and static components with sliding mesh, AMI.

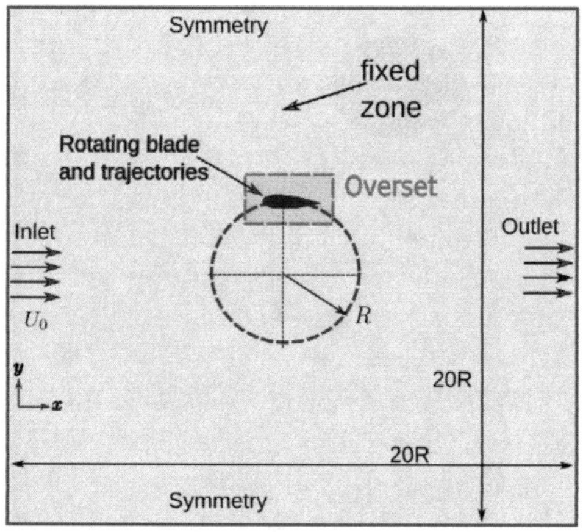

Figure 5: Schematic diagram, computational domain, and boundary conditions of a VAT simulation. Fixed and static components with overset mesh, overset.

The inlet boundary employs a Dirichlet (fixed value) condition for velocity and a Neumann (zero-gradient) condition for pressure. Conversely, the outlet utilizes a Dirichlet condition for pressure and a Neumann condition for velocity. Top and bottom boundaries are treated with symmetry conditions, which mathematically translates to zero normal gradient of velocity (U). For stationary walls, a no-slip boundary condition is enforced. For moving walls like the Vertical Axis Turbine (VAT) surface, a moving wall velocity boundary condition is applied. Finally, the Arbitrary Mesh Interface (`AMI`) technique facilitates data interpolation across the interface boundaries between stationary and moving mesh regions. This technique utilizes a cyclic approach

IOP Conf. Series: Materials Science and Engineering **1312** (2024) 012010 doi:10.1088/1757-899X/1312/1/012010

(`cyclicAMI`) for these interfaces. A detailed summary of the boundary conditions is provided in Table 2.

Table 2: Parameters of the boundary conditions.

Boundary	**Condition**	**Value**
Inlet	Velocity Inlet	$U_0 = 0.183$ m/s, and turbulence
Outlet	Pressure Outlet	$P = 0$ Pa
TopAndBottom	zeroGradient	extrapolated
FrontAndBack	empty	empty
Blade	movingWallVelocity	kqRWallFunction, nutkWallFunction, epsilonWallFunction, omegaWallFunction
Overset	overset	interpolated

3.4. Mesh generation

Two types of meshes are created for the simulations: one for the sliding mesh and another for the overset mesh. In both instances, the mesh exhibits finer resolution near the center and coarser resolution towards the outer region. The coarser mesh is constructed in the background, comprising structured rectangular cells, and refined through various refinement zones. Figure 6 and Figure 7 demonstrate these procedures for SM and OM, respectively.

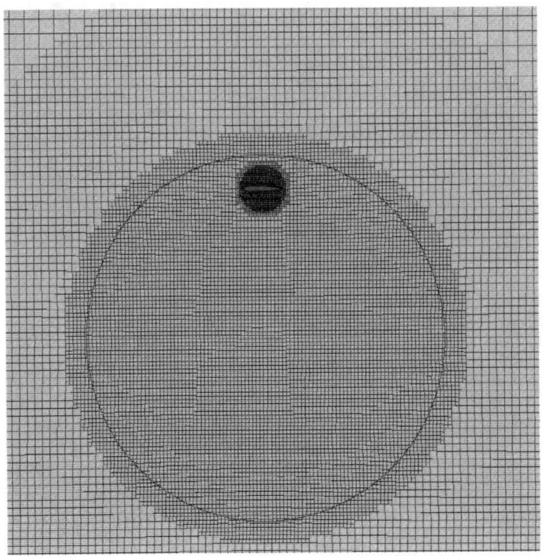

Figure 6: Mesh Generation using SM with AMI.

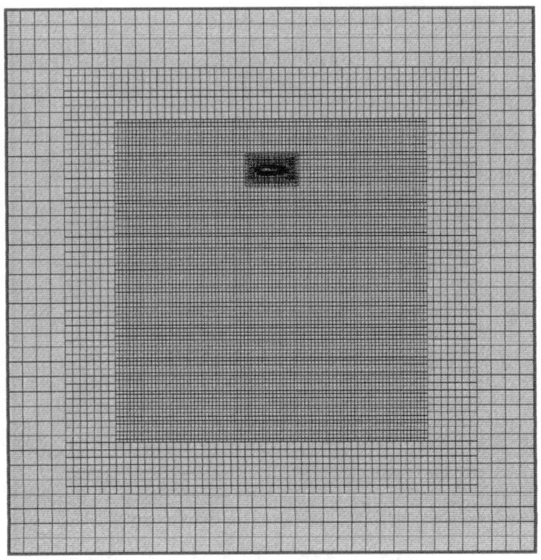

Figure 7: Mesh Generation using OM with overset.

The mesh is constructed using computational tools provided by OpenFOAM, an open-source computational fluid dynamics software package. We adhere to established guidelines for assessing mesh quality, given the absence of a universally accepted metric. Mesh refinement is determined through grid independence tests, ensuring computational results are not unduly influenced by mesh resolution. Additionally, we evaluate key mesh quality parameters such as orthogonality, skewness, aspect ratio, and smoothness, as recommended in prior studies [26]. Refinement near the airfoil's trailing edge is achieved using specialized settings within the `snappyHexMesh` utility, a feature within OpenFOAM, Table 3. The non-dimensional wall distance, denoted as y^+, is utilized to precisely position the first grid cell adjacent to the airfoil surface. This turbulence parameter is defined as,

$$y^+ = \frac{U_f \times y_\perp}{\nu}, \tag{7}$$

where y^+, U_f, $y_\perp$, and ν are the non-dimensional wall distance, friction velocity, the first cell center distance, and the fluid kinematic viscosity, respectively. The friction velocity is obtained using $U_f = \sqrt{\tau_w/\rho}$. Here, $\tau_w = 0.5\rho {U_0}^2 C_f$ is the wall shear stress and C_f local skin friction coefficient [27]. With wall functions, we can use higher values of $y^+ \sim (10 - 30)$ [28].

The fluctuation of y^+ is driven by the continuous rotation of the VAT blade, leading to alterations in the blade's relative velocity. Notably, the area with the highest relative velocity, typically situated close to the azimuth position of 0 degrees, exerts a significant influence on y^+.

Table 3: Main snappyHexMesh parameters and mesh statistics.

snappyHexMesh parameters	**Values**	**Mesh statistics**	**Values**
resolveFeatureAngle	15	Layers	10
nSolveIter	100	Expansion ratio	1.25
finalLayerThickness	0.7	Total cells	60,446
minThickness	1.16×10^{-4}	Rotor cells	31,320
maxNonOrtho	180	Background cells	29,126
maxBoundarySkewness	-1 (disabled)	Max aspect ratio	4.3576, (OK)
maxInternalSkewness	-1 (disabled)	Max skewness	1.4490, (OK)

Figure 8 show the mesh generation with a successively enlarged view near the airfoil. we first studied the convergence of the tangential force coefficient for the solution sensitivity to gird size, number of revolutions, and time step. The grid that improved the overall accuracy is considered for all simulations, 60,000 cells in this case. The computational domain and boundary conditions are shown in Figure 4.

3.5. Mesh generation: Overset mesh

The traditional moving mesh techniques (sliding meshes, morphing meshes, etc) can simulate moderately challenging problems. However, problems involving complex motions such as those that can be described using `6DOF` or with prescribed motion tables are impossible to simulate using these methods. The Overset Meshes OM or (overlapping grids) is a mesh generation strategy that allows a set of cells (control

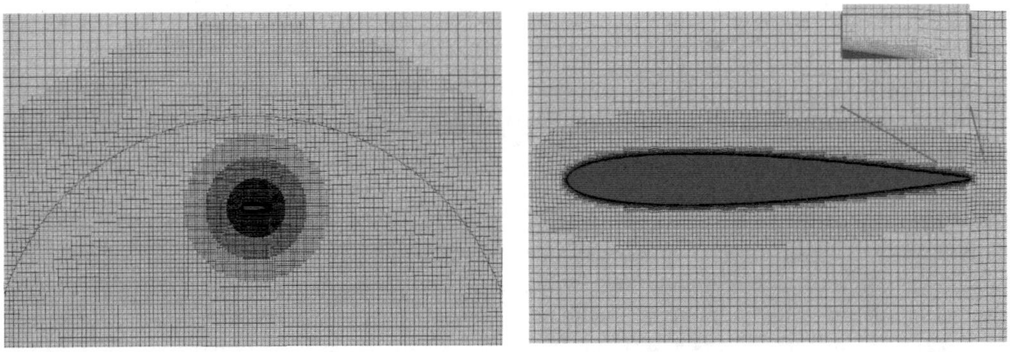

Figure 8: Mesh generation around the airfoil (SM), min edge length of 0.00025 and 10 layers with ER of 1.15. (a) enlarged view close to the airfoil, (b) enlarged view around the airfoil surface.

volumes) to overlap with other sets of grids. OM method enables the simulation of problems that are extremely difficult for traditional moving mesh methods. OM is recognized as an attractive approach for treating complex geometries, single or multiple bodies performing complex movements, allowing great flexibility in modeling them. It is only recently supported by the ESI version of the major open-source software package, OpenFOAM [19, 14, 27].

3.6. *Turbulence model specification*

Unsteady Reynolds-Averaged Navier-Stokes (URANS) simulations were employed to predict the performance of a Darrieus wind turbine in incompressible flow. We initially evaluated three turbulence models: laminar, SST $k-\omega$, and RNG $k-\epsilon$. The laminar model exhibited oscillatory behaviour and overestimated blade loads, rendering it unsuitable for this application. Conversely, both SST $k-\omega$ and RNG $k-\epsilon$ models produced closely matching and accurate results for the tangential aerodynamic forces. The values for turbulent kinetic energy (k), dissipation rate (ϵ), and specific dissipation rate (ω) are estimated using Eq. 8. We set a 5% turbulent intensity (I) and equate the length scale (l) to the chord length (c). The prescribed boundary conditions require specification of the value of U, k, ϵ and ω.

$$k = \frac{3}{2}(UI)^2, \qquad \epsilon = C_\mu^{0.75} \times \frac{k^{1.5}}{l} \qquad \omega = \frac{\epsilon}{k}, \qquad \nu_t = 0.09\frac{k^2}{\epsilon}, \tag{8}$$

The RNG $k-\epsilon$ (`RNGkEpsilon` in OpenFOAM) model is a commonly used turbulence model suitable for high Reynolds number simulations. It does not directly resolve the viscous sublayer near walls. In OpenFOAM, wall modeling with `RNGkEpsilon` is facilitated by the `kqRWallFunction`, which represents turbulence near walls. For Reynolds number around 67000, a y^+ value between 30 to 100 is computationally efficient. However, it's crucial to ensure a sufficiently refined mesh near walls for accurate results. A y^+ value of approximately 8, although more computationally demanding, is

IOP Conf. Series: Materials Science and Engineering **1312** (2024) 012010 doi:10.1088/1757-899X/1312/1/012010

generally acceptable, capturing the boundary layer's logarithmic region while relying on wall functions to handle the viscous sublayer.

The SST $k-\omega$ model is a widely used two-equation turbulence model that blends the $k-\omega$ formulation in the boundary layer with the $k-\epsilon$ behavior in the free stream [29, 30]. This enables accurate prediction of flows with adverse pressure gradients and separation. Unlike the RNG $k-\epsilon$ model, SST $k-\omega$ directly resolves the viscous sublayer near walls, eliminating the need for wall functions. However, it requires a finer mesh resolution near walls compared to RNG $k-\epsilon$ when using wall functions.

The simulations described in this study utilize OpenFOAM-v9 for the sliding mesh (SM) simulations and ESI OpenFOAM-v2112 for the overset mesh (OM) simulations. The solvers employed are `pimpleMFoam` for the sliding mesh and `OverpimpleDyMFoam` for the overset mesh. The PIMPLE solver, integrating the steady-state SIMPLE [31] and transient PISO methods, adopts a pseudo-transient approach. This hybrid algorithm is particularly suitable for turbomachinery-type problems, providing faster simulations compared to fully transient methods, while maintaining higher accuracy than steady-state solvers.

4. Results and discussion

This section presents the results and analysis of 2D numerical simulations conducted to investigate the performance of a VAT utilizing sliding and overset mesh techniques. We use `pimpleMFoam` with out LTS and `OverpimpleDyMFoam` with LTS for SM and OM respectively. LTS is a versatile technique that can be applied to various types of simulations (steady-state and unsteady) to improve computational efficiency and accuracy. Initial sensitivity and convergence analyses were performed using sliding mesh, followed by validation against experimental data. Next, simulations are conducted using the overset mesh and the results are compared. Due to the availability of reference data for validation and its good performance, all the computations employ the symmetric NACA 0015 airfoil.

For the simulations with sliding mesh (SM), we employ a lower `CFL` number (0.9) and this resulted in stability for the complex turbine motion. The SM simulations apply single time step throughout the domain and do not typically utilize local time stepping (LTS). The average value of y^+ is approximately 5, and backward difference is employed for temporal discretization.

For the simulations with overset mesh (OM), we employ a higher `CFL` number (2.0) was chosen to expedite convergence. We also use local time stepping (LTS) which allows different time steps in different regions of the mesh to improve efficiency for complex geometries. Both cases use the RNGkEpsilon turbulence model with `kqRWallFunction`. The average value of y^+ is around 10, and the `Euler` scheme is employed for temporal discretization.

4.1. Aerodynamic Performance

Figure 9a to Figure 9d illustrate the convergence of tangential force coefficient and the solution sensitivity to gird size, number of revolutions, and time step. The grid that improved the overall accuracy is considered for all simulations, about $60,000$ cells in this case. Figure 9b depicts the convergence with revolution cycles for dynamic mesh method separately, sliding mesh. From the figure, it can be seen that the sliding mesh does not converge well till the 6th cycle for which the overset mesh converges, and all simulations generally converge after ten rotations. Even though the mesh size is more for the overset case, the simulations are faster.

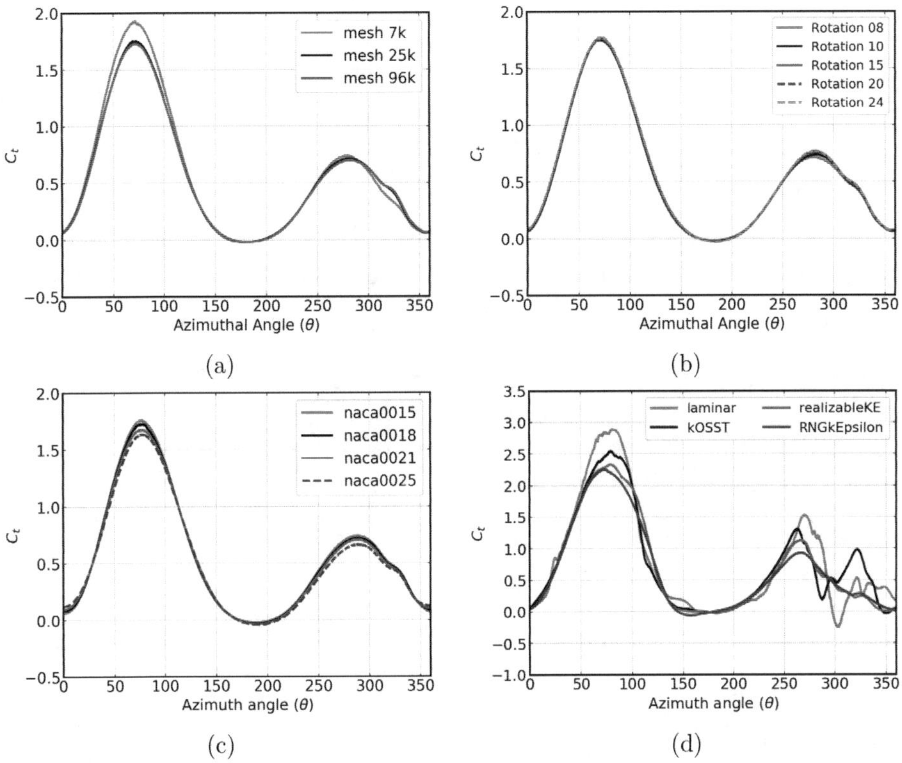

Figure 9: VAT simulation based on SM and sensitivity analysis. (a) mesh sizes, and (b) number of revolutions, (c) blade profiles, and (d) major turbulence models.

Previous studies on VATs have relied on 4-digit symmetric NACA series airfoils, such as NACA0012 [32]. Recent research efforts, however, have ventured into airfoils with higher thicknesses such as the NACA0015, NACA0018, and NACA0021 [32, 33] which offer favourable structural properties, regardless of the associated increase in profile drag. For a comprehensive comparison of their performances, a baseline case with a freestream velocity of 0.183 m/sec and a TSR value of 2.5 has been numerically simulated. Subsequently, various symmetric airfoils were analyzed under this frozen operating condition to discern their performance coefficients.

Figures 9c illustrates the tangential force coefficient on the blade surface of different symmetric blades of thickness ranging from 12 to 25 percent studied at a TSR of 2.5. The airfoil profile with a thinner cross-section performs better (with a higher torque coefficient) at the specified tip speed ratio over the entire rotation, even though the difference between the results of the NACA blades is small. It can be observed that the thinner blade experiences a larger pressure difference between the upper and lower sides of the blade. For the same geometric angle of attack, the thinner blade generates a higher lift force. This result agrees with the study performed by [30] but contradicts the analysis of [34] who computed the flow for a wide range of turbine Reynolds numbers and tip speed ratios. However, [34] used a multiple-streamtube model and also depended on static airfoil data that showed thicker airfoils to outperform the thinner ones.

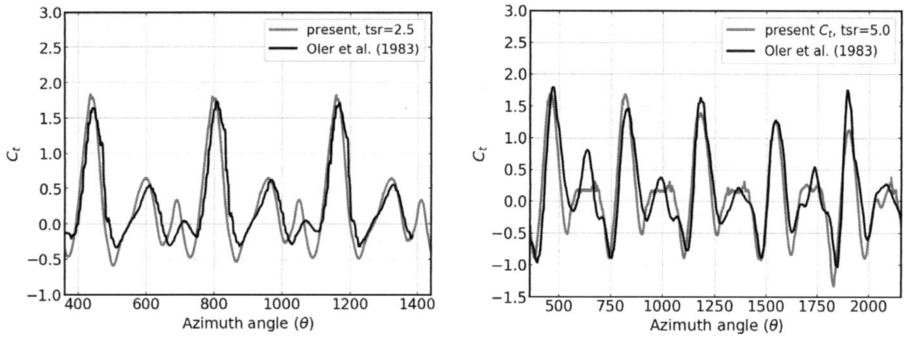

Figure 10: Comparison of tangential force coefficient of SM with the experiments of Oler et. al., [24], (a) $tsr = 2.5$, and (b) $tsr = 5.0$.

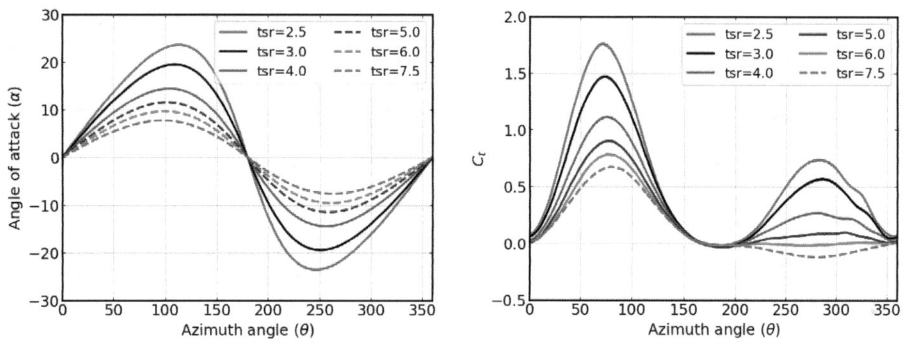

Figure 11: (a) Flow incidence angle (α) versus the blade azimuth position for different tip speed ratios λ, (b) variation of tangential force coefficient versus the blade position for different tip speed ratios λ.

In the upwind pass of Figures 10a, the power extracted is very high while in the downwind pass the speed is highly reduced and the simulated tangential, and normal forces are very low. The simulation is in good agreement with the general pattern of the experimental data. However, the simulation show over prediction at around $\theta = 90°$ and under predicted at around $\theta = 270°$. Even though our simulation conditions are identical, the meshing strategy used is different. One reason for the small discrepancy may be because [24] used a low aspect ratio blade for his experiment that might indicate the presence of 3D effect. Another reason, for the over-predicted simulation is due to the consideration of low Reynolds number which decreases the lift and increases the drag of the blade, [1, 35].

Figures 11 presents the simulation for different tip speed ratios (λ). Figures 11a shows the flow incidence angle (α) versus azimuth position (θ) of the blade for different λ. Figures 11b depicts the variation of C_t for different tip speed ratios (λ).

The overset mesh technique is initially evaluated using different Reynolds numbers to ascertain its accuracy and efficiency. This evaluation entails investigating different flow configurations and complexities with the objective to validate the capability of overset meshes in accurately simulating flows with low to intermediate Reynolds numbers.

Figure 12, depicts the coefficient of lift for a flapping airfoil subjected to heaving and pitching motions at Reynolds number $Re = 1000$. The results demonstrate satisfactory agreement with the reference data [36], effectively capturing the essential features of the lift curve. Additionally, Figure 13 presents the lift coefficient for a pure plunging airfoil at a slightly higher Reynolds number, $Re = 4 \times 10^4$, showing good agreement with the reference data [37].

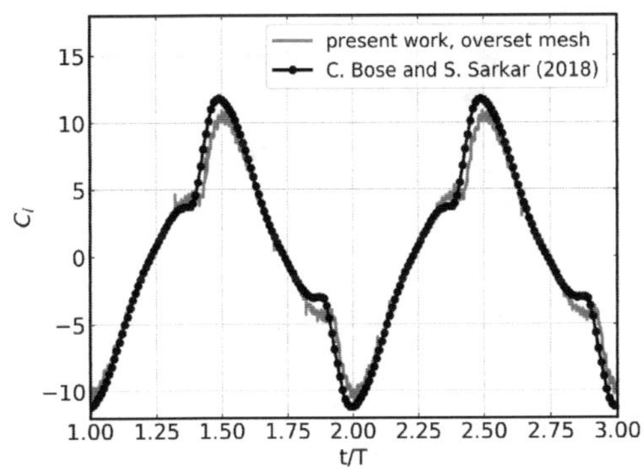

Figure 12: Pitching-plunging airfoil at $Re = 1000$, comparison of the lift coefficient C_l with [36] in the third revolution.

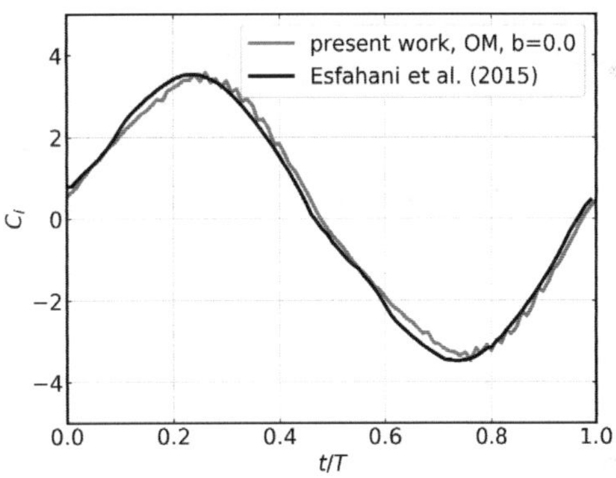

Figure 13: Pure-plunging airfoil at $Re = 4 \times 10^4$, comparison of the lift coefficient C_l with [37] in the fifth cycle.

Figure 14 compares the simulated results of a single-blade VAT based on the sliding and the overset meshes [19]. Both approaches predict the tangential force coefficient very well with the overset mesh showing minor discrepancies, however, all major features of the curve match satisfactorily. The capability of the overset mesh method to utilize `tabulated6DoF` Motion makes it more convenient for simulating more complex motions.

While the sliding mesh is a typical modeling technique for wind and hydrokinetic turbines, the overset method is also rarely used for such applications. However, there is little or no standard guideline for the design, simulations, and analysis of such systems using CFD in the literature and the available best practices differ from one to the other. The current work presents the methodology for accurate simulation of a VAT employing the above approaches. Single to multiple-bladed turbines can now be analyzed using

IOP Conf. Series: Materials Science and Engineering **1312** (2024) 012010 doi:10.1088/1757-899X/1312/1/012010

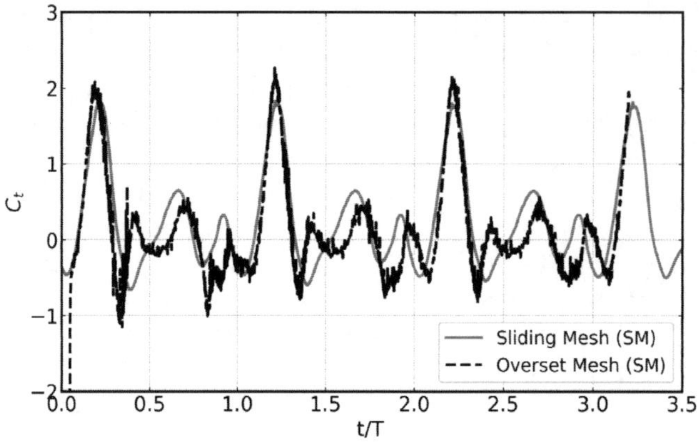

Figure 14: Comparison of the tangential force coefficient of VAT simulation based on sliding and overset meshes, over three cycles.

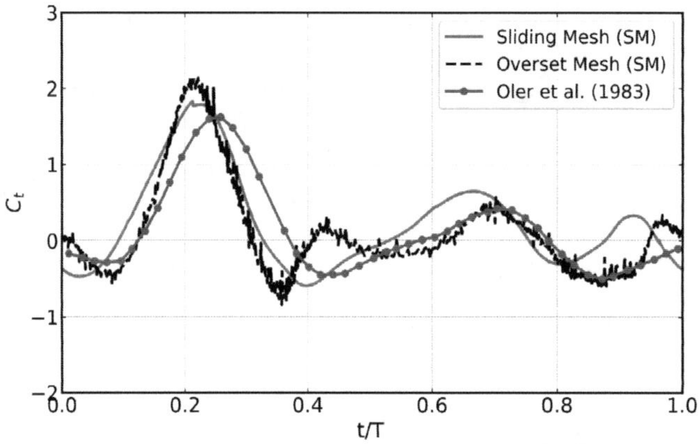

Figure 15: Comparison of the tangential force coefficients of sliding mesh, overset meshes, with experiment over one cycles.

OpenFOAM CFD with a lesser number of cells. For a single-blade simulation, the SM and OM predicted all the major features of the torque curve very well and with almost similar accuracy. The comparison of these dynamic mesh techniques by [14] reported that the peak of the torque curve is predicted at different TSRs and that the OM method converges faster.

4.2. Flow Field Simulation

Figure 16 shows snapshots of the vorticity distribution and the associated mesh during two periods of turbine revolution at four different time steps. The vorticity distribution shows a similar pattern for the sliding mesh and overset mesh cases. The left row indicates the mesh motion. The snapshots in the middle and right rows indicate the results obtained using the sliding and overset meshes, respectively.

Figure 2, the optimal performance for a VAT occurs when the blade aligns near the perpendicular position (90°) to the incoming flow in the upwind, as confirmed by [38, 39]. The flow velocity is optimal at that position where the blade sees an effective angle. The track of the turbine blade can be described in four segments: `windward`, `upwind`, `leeward`, and `downwind`. Different aerodynamic feature occurs in each segment. As depicted in Figure 15, the torque profile is least favorable in the leeward segment, where the airfoil aligns with the flow direction. Similarly, the downwind side of the windward segment exhibits notably low torque. Further exploration into the aerodynamics of the blade on an altered trajectory is provided in subsequent sections.

Figure 16 is a qualitative visualization of the vorticity field when the blade is at different positions of the curved and straight line paths during its motion. The blade aerodynamics is more complex around the leeward segment, and the observed behavior can be discussed as:

`Leeward` segment: The incoming flow velocity exceeds the blade velocity in this region (for the simulated TSR of 2.5), leading to a stronger interaction between the blades and the wake. This generates vorticity due to flow separation and mixing.

Lower left end: As the blade reverses its direction and enters the leeward segment, it encounters the previously shed vortices. This sudden change in direction combined with the existing wake leads to significant flow separation and vorticity production in this region.

Lower right end: The shed vortices from previous cycles reach the lower right end before the blades arrive. This interaction between the faster-moving vortices and the incoming flow creates additional vorticity and complex flow patterns.

5. Strengths and Limitations

5.1. Sliding mesh technique:

The sliding mesh method involves moving or deforming the mesh to accommodate the motion of the domain. The mesh is updated at each time step based on the motion of the domain. It has been extensively studied and improved over the years, and its strengths and weaknesses are well known.

The Sliding mesh has limitations in that it only supports a single form of mesh motion or a single topological change within a simulation. This limitation arises because SM depends on the dynamic mesh functionality of the AMI technique, which also requires the geometry to be a symmetrical kind. However, this weakness is replaced in OpenFOAM version 10 which the developers have redesigned it using the non-conformal coupling (NCC) technique [40].

Strengths:

- The sliding mesh is computationally efficient since the mesh is updated locally around the moving objects, rather than globally over the entire domain. This can significantly reduce computational costs and enable faster simulations.
- The sliding mesh technique is suitable to model the rotation of wind turbines. Recent studies show that it can be applied for moving cases with multi-degree-of-freedom such as the combined motions of an underwater vehicles.

Weaknesses:

- SM can lead to mesh distortion, which can affect the accuracy and stability of the simulation.
- SM can be challenging to implement for complex geometries or complex motions.
- SM can result in poor-quality mesh elements, especially near moving objects, which can affect the accuracy of the simulation.

5.2. Overset mesh technique:
The overset mesh method works by creating separate meshes for the moving and stationary domains for the convenience of operating motion functions. A composite mesh is created by patching together the individual meshes using the `mergeMeshes` utility. The overset grid approach offers the convenience of mesh generation and the capability to produce comparable results for complex systems without the drawbacks of mesh deformation. This approach combined with its flexibility proves valuable in exploring ingenious configurations that aim to improve the aerodynamic efficiency of Vertical Axis Turbines. A few examples of clever turbine designs include the dual-axis VAT proposed by [41], the variable geometry oval trajectory VAT concept introduced by [42], and investigations involving variable pitch scenarios as documented in [13]. Moreover, when combined with dynamic mesh techniques like mesh morphing, it becomes possible to shape the blade in specific regions for enhanced performance. The overset mesh technique has undergone substantial research and development, and both its advantages and disadvantages have been well known.

Strengths:

- Overset meshes can handle complex motions and geometries efficiently and excels very well, the SM technique struggles in such cases. Examples are, two cylinders moving toward each other with large relative displacements, track-based vertical axis turbine motion, and flapping of airfoils around an elliptic trajectory.
- It has proven capability of handling multi-body motions, such as the flapping of multiple wings, or the motion of multiple vehicles moving in close proximity to one another.
- It maintains the mesh quality close to moving objects which enhances simulation stability and accuracy. It outperforms the sliding mesh technique for such cases.
- It can be used for modeling multiple moving objects with diverse motions, which is hard to do using the sliding mesh approach and other capabilities in OpenFOAM.

Weaknesses:

- The interpolation methods for overset are not mas conservative [43]. To minimize the error, the background and the overset meshes should have similar cell sizes near the interface. It can also add numerical diffusion to the solution.
- Can be computationally expensive. When multiple moving meshes are used, all need to be patched together and this can slow the simulation by unpredictable factor.
- Rare discontinuities at the interface between moving and stationary geometries.

IOP Conf. Series: Materials Science and Engineering **1312** (2024) 012010 doi:10.1088/1757-899X/1312/1/012010

6. Conclusions

An appropriate meshing technique for Vertical Axis Turbine (VAT) simulations is crucial due to the complex flow field surrounding the moving blades. This study evaluated the performance of two common methods: Sliding Mesh (SM) and Overset Mesh (OM). Both approaches were applied to a baseline VAT case and validated against established benchmarks.

The findings suggest that sliding mesh approach offers a computationally efficient solution for VAT simulations with simple geometries and rigid body motions. It's particularly suited for baseline VAT analysis, especially at higher speed ratios. However, its limitations in handling complex motions and blade modifications restrict its applicability for innovative designs.

Overset mesh excels at simulating VATs with complex kinematics, including morphing blades and localized modifications. This enables targeted performance optimization in specific turbine regions. Additionally, OM maintains good mesh quality near moving objects, a crucial advantage for complex scenarios. While potential drawbacks include non-conservative interpolation methods and increased computational cost, the flexibility and accuracy of OM make it a valuable tool for analyzing advanced VAT designs.

Acknowledgments

We thank the Aerospace Computational Engine (ACE) at the Department of Aerospace Engineering, Indian Institute of Technology Bombay for providing computational resources. Additionally, we thank the organizing committee of the 18th OpenFOAM Workshop for their invaluable support. The first author acknowledges with gratitude the support received from (1) the Indian Institute of Technology Bombay, (2) Wolf Dynamics OpenFOAM training, (3) sponsors of the 18th OpenFOAM Workshop, and (4) organizers of the 18th OpenFOAM Workshop, all of which significantly contributed to the success of this work.

References

[1] Hansen J T, Mahak M and Tzanakis I 2021 *Renewable Energy* **171** 1371–1381
[2] Bianchini A, Balduzzi F, Bachant P, Ferrara G and Ferrari L 2017 *Energy Conversion and Management* **136** 318–328
[3] McCollum D, Gomez Echeverri L, Riahi K and Parkinson S 2017
[4] Sane S, Sudhakar K, Fernandes B and More D 2006 *Advances in Energy Research*
[5] Dabiri J O 2011 *Journal of renewable and sustainable energy* **3** 043104
[6] Ghasemian M, Ashrafi Z N and Sedaghat A 2017 *Energy Conversion and Management* **149** 87–100
[7] Hand B, Kelly G and Cashman A 2021 *Renewable and Sustainable Energy Reviews* **139** 110699
[8] Zhao Z, Wang D, Wang T, Shen W, Liu H and Chen M 2022 *Sustainable Energy Technologies and Assessments* **49** 101789
[9] Bhutta M M A, Hayat N, Farooq A U, Ali Z, Jamil S R and Hussain Z 2012 *Renewable and Sustainable Energy Reviews* **16** 1926–1939
[10] Kumar R, Raahemifar K and Fung A S 2018 *Renewable and Sustainable Energy Reviews* **89** 281–291
[11] Karmakar S D and Chattopadhyay H 2022 *Sustainable Energy Technologies and Assessments* **53** 102469
[12] Balduzzi F, Bianchini A, Maleci R, Ferrara G and Ferrari L 2016 *Renewable Energy* **85** 419–435
[13] Minetto R A L and Paraschivoiu M 2020 *Energy* **202** 117705
[14] Lopez Mejia O D, Mejia O E, Escorcia K M, Suarez F and Laín S 2021 *Processes* **9** 1933
[15] Baghdadi M, Elkoush S, Akle B and Elkhoury M 2020 *Renewable Energy* **154** 239–251

[16] Weller H G, Tabor G, Jasak H and Fureby C 1998 *Computers in physics* **12** 620–631
[17] Windt C, Davidson J and Ringwood J V 2018 *Renewable and Sustainable Energy Reviews* **93** 610–630
[18] Fertahi S e D, Belhadad T, Kanna A, Samaouali A, Kadiri I and Benini E 2023 *Fluids* **8** 242
[19] Guerrero J 2006 *DICAT, University of Genoa, Italy*
[20] Holzmann T 2017 Mathematics, numerics, derivations and openfoam (r), holzmann cfd
[21] Ferziger J H and Perić M 2002 *Computational methods for fluid dynamics* vol 3 (Springer)
[22] Maric T, Höpken J and Mooney K The openfoam technology primer (sourceflux, duisburg, 2014)
[23] Jasak H and Tuković Ž 2010 *Proceedings of the V European Conference on Computational Fluid Dynamics ECCOMAS CFD 2010*
[24] Oler J, Strickland J H, Im B and Graham G 1983 Dynamic-stall regulation of the darrieus turbine Tech. rep. Texas Tech Univ., Lubbock (USA). Dept. of Mechanical Engineering
[25] Inc O Onshape: 3d cad software for everyone online software URL https://www.onshape.com/
[26] Aqilah F, Islam M, Juretic F, Guerrero J, Wood D and Ani F N 2018 *IIUM Engineering Journal* **19** 203–212
[27] Laws P, Saini J S and Kumar A 2019
[28] Nordanger K, Holdahl R, Kvarving A M, Rasheed A and Kvamsdal T 2015 *Computer Methods in Applied Mechanics and Engineering* **284** 664–688 ISSN 0045-7825 isogeometric Analysis Special Issue URL https://www.sciencedirect.com/science/article/pii/S0045782514004034
[29] Rogowski K 2019 *Energies* **12** 2506
[30] Danao L A, Qin N and Howell R 2012 *Proceedings of the Institution of Mechanical Engineers, Part A: J. of Power and Energy* **226** 867–881
[31] Patankar S V and Spalding D B 1983 *A calculation procedure for heat, mass and momentum transfer in three-dimensional parabolic flows* (Pergamon: Elsevier)
[32] Hansen M 2008 *Sterling, VA*
[33] Durrani N, Hameed H, Rahman H and Chaudhry S 2011 *49th AIAA aerospace sciences meeting including the new horizons forum and aerospace exposition* p 541
[34] Healy J 1978 *Wind Engineering* 1–9
[35] Chong W, Fazlizan A, Poh S, Pan K, Hew W and Hsiao F 2013 *Applied Energy* **112** 601–609
[36] Bose C, Gupta S and Sarkar S 2021 *Journal of Fluid Mechanics* **911** A31
[37] Esfahani J, Barati E and Karbasian H R 2015 *Computers & Fluids* **108** 142–155
[38] Porté-Agel F, Bastankhah M and Shamsoddin S 2020 *Boundary-Layer Meteorology* **174** 1–59
[39] Paraschivoiu I, Ammar S and Saeed F 2018 *Transactions of the Canadian Society for Mechanical Engineering* **42** 393–403
[40] Greenshields C and Weller H 2022 *Notes on Computational Fluid Dynamics: General Principles* (Reading, UK: CFD Direct Ltd)
[41] Naccache G and Paraschivoiu M 2017 *J. of Fluids Engineering* **139**
[42] Ponta F, Seminara J and Otero A 2007 *Renewable energy* **32** 35–56
[43] Alletto M 2022 *OpenFOAM® Journal* **2** 13–30

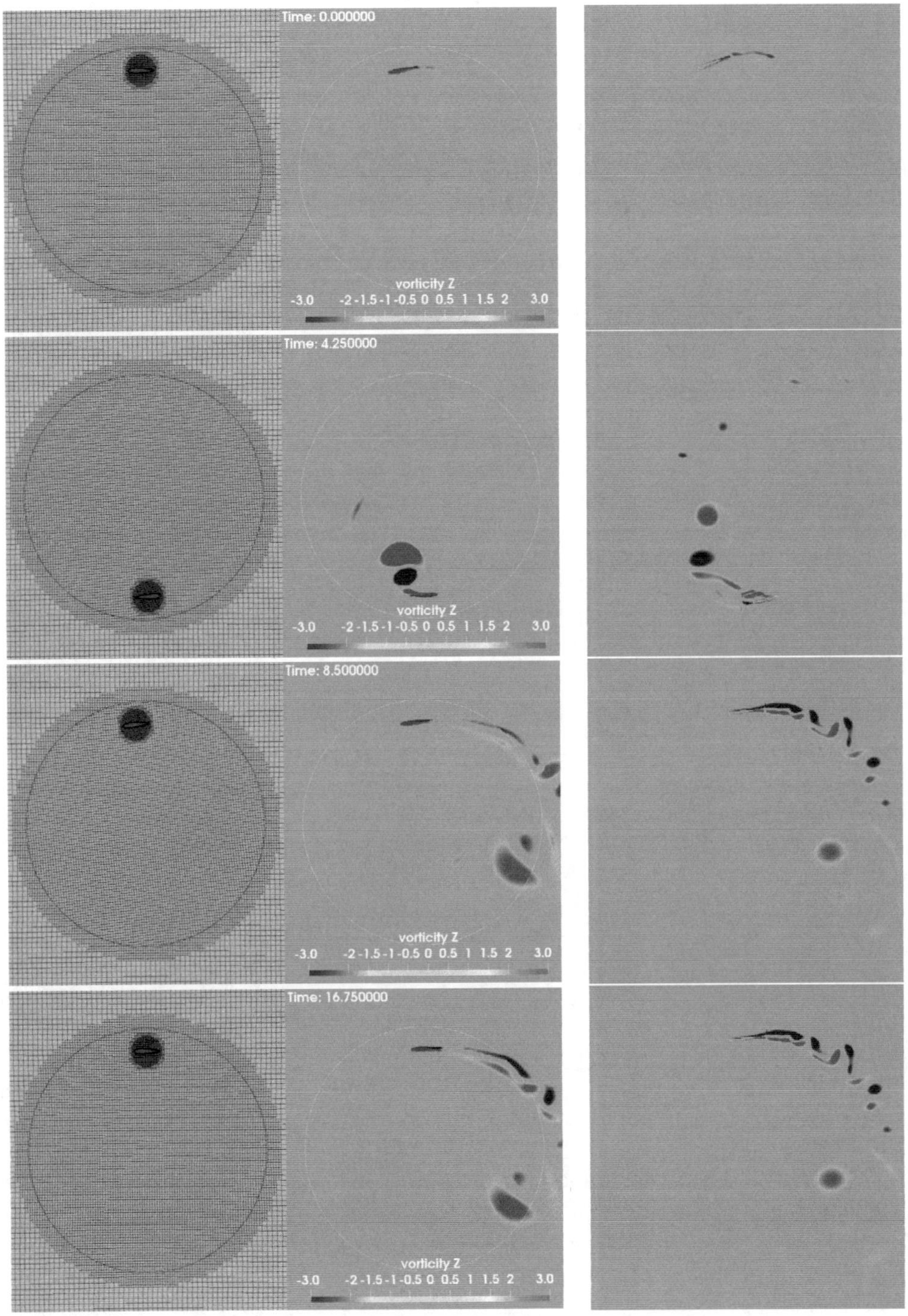

Figure 16: VAT simulation based on sliding and overset meshes, mesh visualization (left) and vorticity distribution over two complete cycles; SM (middle) and OM (right).

IOP Conf. Series: Materials Science and Engineering **1312** (2024) 012011 doi:10.1088/1757-899X/1312/1/012011

Lagrangian properties of the blood flow through human and murine aortic arches: towards improved customised therapies and diagnostic techniques

Simone Ghiglia, Marco Mazzuoli and Joel Guerrero

Dep. of Civil, Chemical and Environmental Engineering (DICCA), University of Genoa, Via Montallegro 1, 16145, Genoa, Italy

E-mail: `marco.mazzuoli@unige.it`

Abstract. The mass transport properties of the blood flow in the aortic arch are investigated by means of direct numerical simulations. The aortic arches in human and murine cases are assumed to be similar (with similarity factor equal to 20) and characterised by constant diameter, while the axis lays on a plane. The simulations were carried out using OpenFOAM (v.10). The flow regime appears remarkably different in the two cases because flow separation and vortical structures appear during the systolic phases in the human case, which are absent at the mouse scale. Consequently, peaks of the wall-shear-stress occur at different phases and, in the murine case, are characterised by a magnitude nearly 4 times larger than in the human case. The trajectories of fluid particles are computed in order to evaluate the dispersion efficiency exploited by biomedical applications (e.g. drug delivery or solid micro carriers). Despite the different flow regimes, in both system particles uniformly released at the inlet preserve a homogeneous distribution as they flow in the aortic arch. In particular, during the early decelerating phases of systole, the fluid trajectories are found frequently to approach the zones of the wall where the shear-stress is large.

1. Introduction

Cardiovascular diseases (CVDs) represent one the major causes of death all over the world. According to World Health Organization (2019) data, around 17.9 million people die every year from CVDs, which correspond to around 32 percent of annual worldwide deaths. Among CVDs, atherosclerosis is one of the most common morbidities that afflict the elderly population and its early diagnosis is of primary importance in order to administer appropriate treatments to patients and prevent premature deaths. With the aim of reducing the impact of these pathologies, innovative research fields of modern vascular medicine concern the exploitation of nanomedicine for the diagnosis and treatment of atherosclerosis [10]. For instance, engineered nanoparticles, hereinafter referred to as nanocarriers (NCs), capable of selectively recognise and bind the pathological sites, can be used to improve techniques [3, 19, 7, 6, 5] (cf.figure 1). A critical aspect of such technique lays in the event that certain potential pathological regions of the arterial vessels may be never reached by NCs which are erratically transported by the blood flow. In the present contribution, the transport properties of blood flow in the aortic arch are studied by means of Direct Numerical Simulation (DNS). The DNSs were performed considering simplified geometries of the aortic arch, which resemble the human and murine

Figure 1. Visualisation of the upper part of the aortic arch in the phase when bulk velocity is maximum (run 2). Vortical structures are identified by surfaces at $\lambda_2^* = 25\ \omega^{*2}$. NCs (red dots) and the terminal part of their trajectory are indicated.

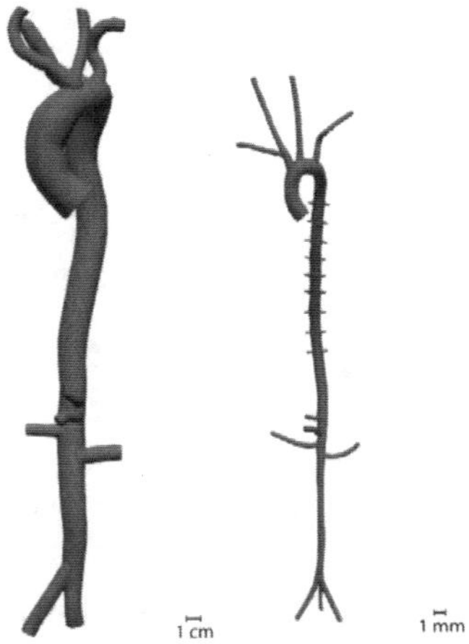

Figure 2. Model of young human (left) and murine (right) aortic arch [9].

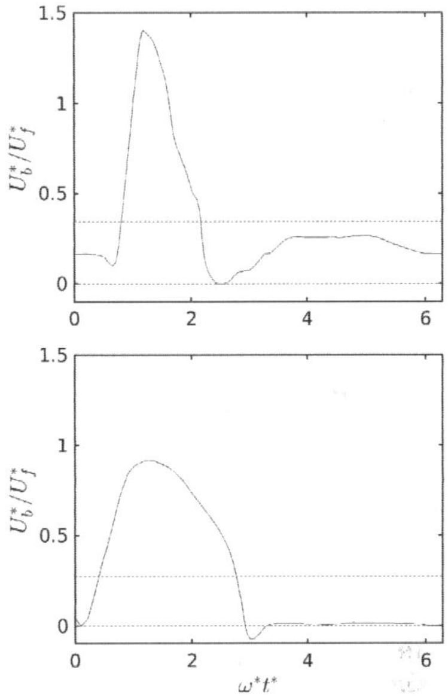

Figure 3. Bulk velocity U_b^* during a pulsation (black curves) and its mean value U_s^* (red lines) normalised by U_f^* for run 1 (top) and run 2 (bottom).

actual arches and were adopted in previous studies [11, 13, 14] (see figure 2). The objective of the numerical investigation is twofold: (i) measuring the capability of NCs to inspect the internal surface of the aortic arch and (ii) identifying the differences of NC dispersion in human and murine cases. Indeed, most of the clinical tests are carried out on mice even though the blood flow properties are remarkably different from those in humans. For instance, as for the development of atherosclerotic plaques is concerned, the aortic arch is critical in mice, whereas in humans atherosclerosis is more commonly diagnosed in the lower limbs. This is conceived as a preliminary study of the Lagrangian transport properties of arterial vessels. In the following sections the methodology and the main results are briefly described.

2. Numerical approach

Two simulations were performed, run 1 and run 2, which mimic the blood flow in the human and murine aortic arches, respectively. The model of aortic arch used for the simulations consists of a rigid tube characterised by a circular cross-section of constant diameter, the axis of which lays on a vertical plane. Blood in such large arteries is assumed to behave as a Newtonian fluid with density ϱ^* and kinematic viscosity ν^*. A finite volume approach is employed, with a hybrid mesh consisting of polyhedral and prism elements. The computational domain in the near-wall region is discretised by 8 layers of stretched grid elements characterised by stretching factor equal to 1.2. Blood pulsating flow is considered (see figure 3) and non-iterative PISO algorithm is used, where the time advancement of the viscous terms is made with a Crank-Nicolson scheme. The vortical structures shed downstream the vessel bends were fully resolved without the aid of turbulence models. The algorithm is implemented in OpenFOAM [18].

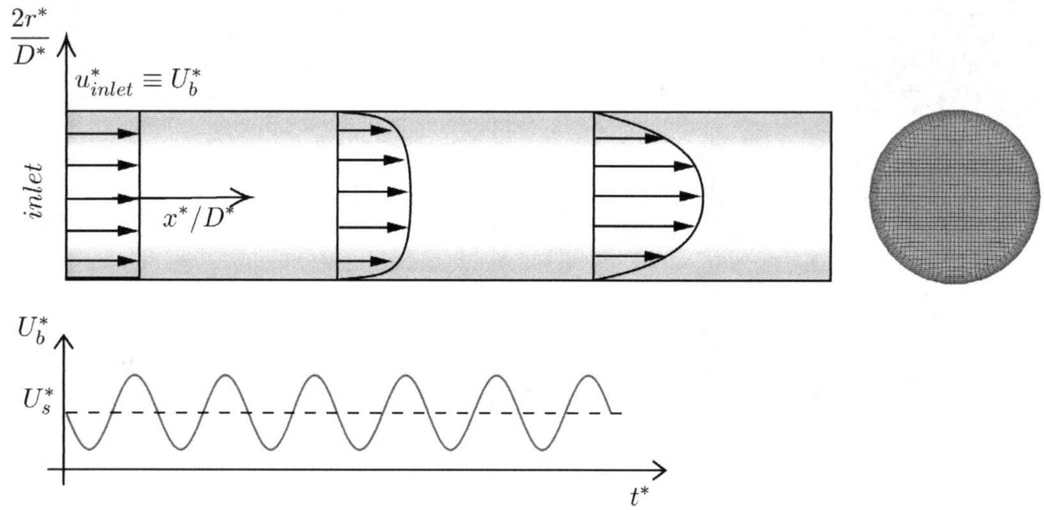

Figure 4. Validation test: sketch of the flow configuration considered by [2] (top) and time development of the bulk velocity U_b^* (bottom). The cross-sectional mesh is shown in the right panel.

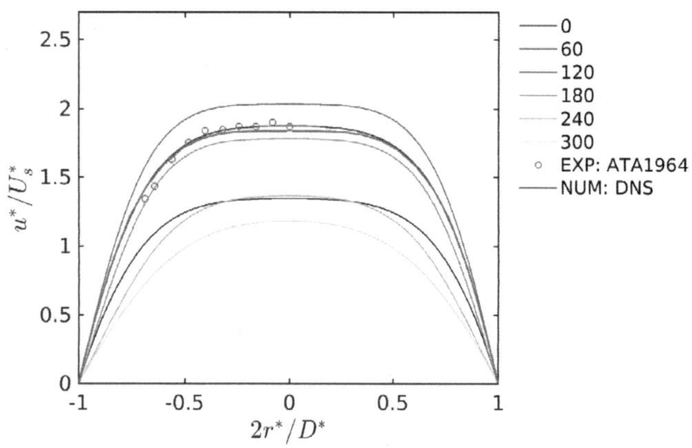

Figure 5. Validation of the numerical approach: comparison of the experimental (red circles) and analytical (grey/blue lines) results obtained by [2] with the numerical results at time $\omega^* t^* = \pi/3$ (red line) at the same distance from the inlet $x^*/D^* = 0.0144\, Re_b$. The values in the legend refer to the pulsation phase $\omega^* t^*$ expressed in degrees.

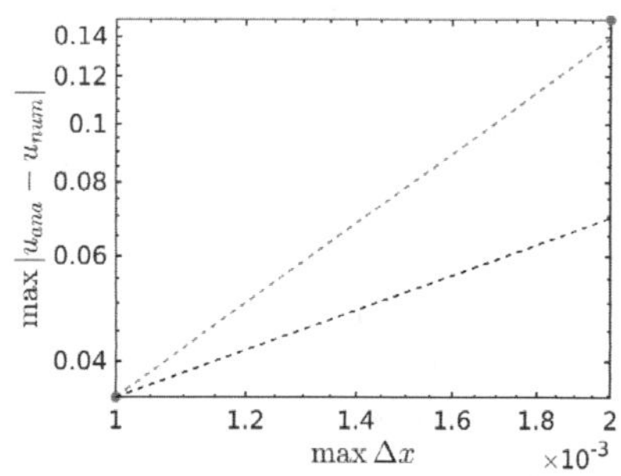

Figure 6. Maximum error of the velocity field with respect to Atabek & Chang's solution [1]. Black and red dashed lines indicate the first-order and second-order trends respectively, while red dots were obtained by numerical simulations for different maximum size of the mesh cells.

2.1. Validation of the numerical approach

The numerical approach was preliminarily validated against the analytical solution and the experimental measurements obtained by [2] for a laminar spatially-developing pulsating flow in a straight circular tube of diameter D^*. Hereinafter, dimensional quantities are marked by the superscript *. The flow configuration is sketched in figure 4 along with the mesh at a cross section and the time development of the bulk velocity. Uniform pulsating unidirectional velocity of magnitude equal to the bulk velocity $U_b^* = 4Q^*/(\pi D^{*2})$ is imposed at the inlet of the tube, Q denoting the blood flow rate. In particular, sinusoidal velocity oscillations are superimposed on a steady stream of magnitude $U_s^* = \overline{U_b^*}$, the overline indicating the time average. Hence, the bulk Reynolds number is defined as

$$Re_b = U_s^* D^*/\nu^* ,$$

which is equal to 1500. Moreover, the problem is characterised by the Womersley number

$$\alpha = D^* \sqrt{\omega^*/\nu^*}$$

and the Reynolds number of the boundary layer

$$Re_\delta = U_f^* \delta^*/\nu^* ,$$

equal to 10 and 310, respectively. The amplitude of velocity oscillations of the developed flow at the tube centre is defined as $U_f^* = \sqrt{8}\ \mathrm{rms}(U_b^* - U_s^*) = 2\,U_s^*\sqrt{a^2+b^2}$, where $U_b^* - U_s^* = U_s^*[a\cos(\omega^* t^*) + b\sin(\omega^* t^*)]$, $a = -0.123$ and $b = 0.321$. The viscous length characterising the oscillatory boundary layer is denoted by $\delta^* = \sqrt{2\nu^*/\omega^*}$, where ω^* is the pulsation frequency. Figure 5 shows a fair comparison between the present numerical simulation and the analytical and experimental values reported by [2], during the accelerating phases at a distance $x^* = 21.6\ D^*$ from the inlet. Finally, the second-order accuracy of the numerical scheme was also verified (cf. figure 6).

2.2. *Geometrical features of the aortic arch in the simulations*

The diameter of the murine and human aortic arches is not uniform and change during the life span [9]. Similarly, the axis of the arch does not lay on a plane and its curvature is not constant (cf. figure 2). Nonetheless, for young humans and mice, these features can be assumed to have moderate effects on the flow dynamics [14]. In particular, the young murine aortic arch is relatively well described by a tube of constant diameter and constant curvature $1/R^*$, R^* denoting the arch radius. Hence, an additional dimensionless parameter is introduced, namely the dimensionless arch radius defined as

$$\mathcal{R} = 2R^*/D^* .$$

The value of $\mathcal{R}$ depends on the age and on the species (e.g., for a human aortic arch, it can range between 2.4 [4] and 3.6 [12]). The numerical approach was further validated comparing the magnitude and the distribution of the wall-shear-stress (WSS) obtained in the present simulations of a human aortic arch with the measurements that Callaghan & Grieve [4] made by means of Magnetic Resonance Imaging (MRI) (cf. figure 7). Moreover, the present WSS distribution fairly matches that numerically obtained by Suito et al. [14] (see figure 13 of [14]).

2.3. *Computation of NC trajectories*

NCs are assumed to be spherical and characterised by the dimensionless diameter $d = d^*/D^*$ and relative density $s = \varrho_s^*/\varrho^*$. A Lagrangian point-particle approach is used to track the individual NCs. The diameter of NCs d^* is much smaller than the diameter of the aortic arch D^* and the particle inertia is approximately negligible. In fact, the particle Stokes number

$$St = \omega^* d^{*2} \varrho_s^*/(18\varrho^* \nu^*)$$

is found to be much smaller than 1, therefore solid particles are assumed to exhibit the same dynamics as fluid particles. It is also assumed that the particles stop as they come into contact with the wall, which is considered the target of NCs for diagnostic purposes. The number of particles presently tracked is of order $\mathcal{O}(10^4)$ for both human and murine cases.

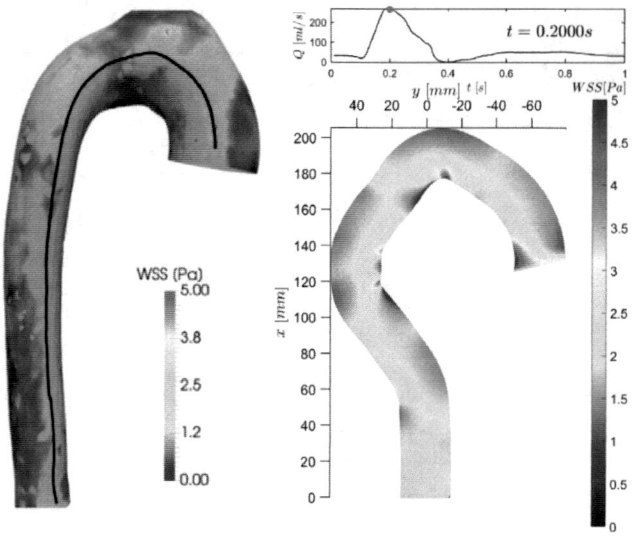

Figure 7. WSS distribution obtained from MRI data of [4] (left panel) and from the present run 1 at the phase when the flow is maximum (right panel). Human aortic arch.

Table 1. Simulation dimensional parameters: vessel diameter D^*, mean bulk velocity U_s^*, amplitude of bulk velocity fluctuations $U_f^* = \sqrt{8}\,\mathrm{rms}(U_b^* - U_s^*)$, pulsation angular frequency ω^*, blood density ϱ^*, NC diameter d^*.

Run #	D^* [mm]	U_s^* [cm/s]	U_f^* [cm/s]	ω^* [s^{-1}]	ϱ^* [g/cm^3]	d^* [nm]
run 1 (human)	30	9.42	27.4	6.28	1.0	40
run 2 (mouse)	1.5	10.2	37.1	44.9	1.0	40

Table 2. Simulation dimensionless parameters: bulk Reynolds number Re_b, Stokes Reynolds number Re_δ, Womersley number α, arch radius $\mathcal{R}$, Dean number $\mathcal{D}$, secondary streaming Reynolds number Re_s, NC relative density s, dimensionless particle diameter d, particle Stokes number St.

Run #	Re_b	Re_δ	α	$\mathcal{R}$	$\mathcal{D}$	Re_s	s	d	St
run 1 (human)	707	77	37.6	3.3	212	1789	19.3	10^{-6}	10^{-9}
run 2 (mouse)	38	39	5.02	3.3	11	454	19.3	10^{-5}	10^{-8}

3. Results and discussion

As aforementioned, two DNSs were carried out, referred to as runs 1 and 2, which mimic the blood flow in human and murine aortic arches, respectively. The values of the dimensional parameters characterising the simulations are indicated in table 1 while those of the dimensionless parameters, which were defined in the previous section, are reported in table 2. Alternatively, relevant dimensionless parameters, that can be obtained by combining the previous ones, are the Dean number and the secondary streaming Reynolds number defined as

$$\mathcal{D} = Re_b/\sqrt{\mathcal{R}}\,, \quad Re_s = Re_\delta^2/2\mathcal{R}\,,$$

respectively. For these values of the parameters, the pulsating flow in a straight tube would be laminar [16], while in the present cases the axial curvature induces secondary motions and may

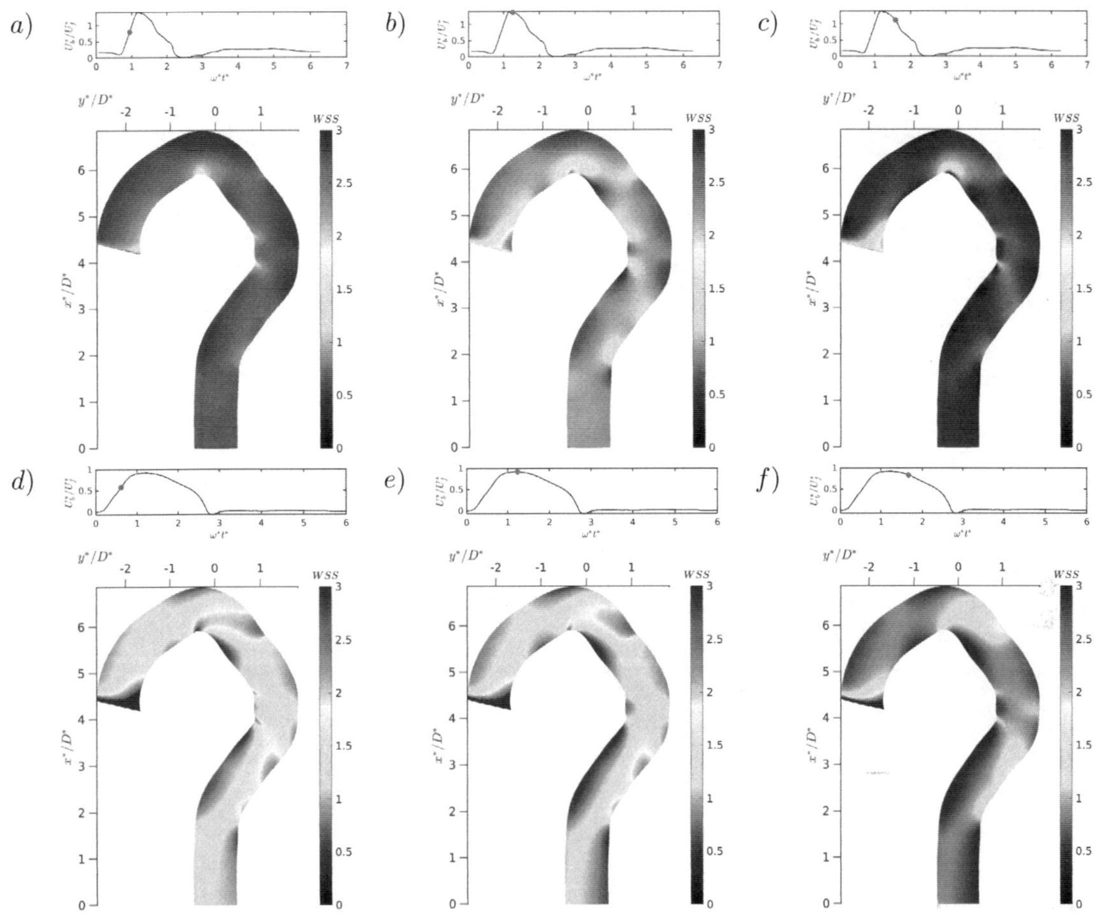

Figure 8. Wall shear stress, normalised by $\varrho^* U_f^* \omega^* \delta^*$, at the peak velocity phase (run 2).

cause the boundary layer separation. The minimum distance of mesh cells from the wall is 0.14 δ^* for run 1 and 0.019 δ^* for run 2. The dimensionless bulk velocities during a cardiac pulsation that were considered for the present simulations are shown in figure 3. The dimensionless diameter of NCs and the Stokes number are small ($d \ll 1$ and $St \ll 1$), and the particle dynamics can be assumed unaffected by the value of the specific gravity $s = \varrho_s^*/\varrho^*$.

The flow properties of runs 1 and 2 are compared and solid particle trajectories are computed to evaluate the effectiveness of the diagnostic technique. Vortices are identified using the λ_2-criterion (cf. figure 1) and the vorticity is analysed. During the deceleration phases and the early diastolic phases, vortical structures form in run 1 as an effect of the boundary layer separation downstream of the vessel bends, which are characterised by the vorticity component orthogonal to the arch plane. The flow in the regions characterised by the presence of these vortices is chaotic. Such vortex structures are not observed in run 2. Moreover, in both runs, vortices are observed in the upper part of the aortic arch throughout the systolic phase of the pulsation, which are characterised by streamwise vorticity, namely they are associated with the secondary flows orthogonal to the mean flow direction (see figure 1). The WSS is investigated since large shear stresses potentially contribute to the development of atherosclerotic lesions, leading to the activation of pro-inflammatory and pro-atherogenic pathways. Figure 8 shows the WSS distribution at different phases of the pulsation. While in run 1 the maximum

IOP Conf. Series: Materials Science and Engineering **1312** (2024) 012011 doi:10.1088/1757-899X/1312/1/012011

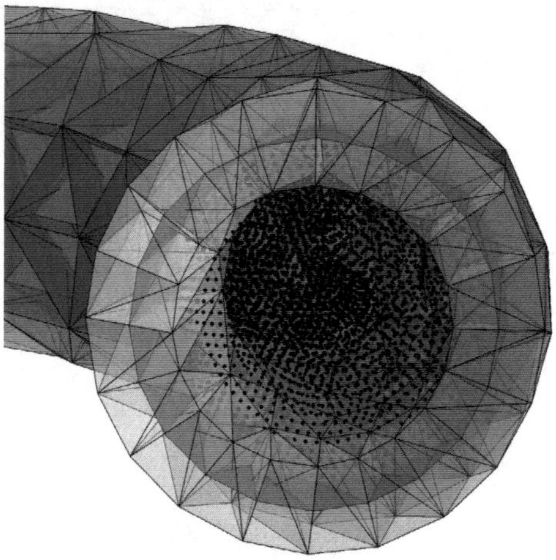

Figure 9. Method to compute the distribution of the NCs approaching the surface of the aortic arch: the vessel surface (in red) is divided in triangles of identical area which are extruded in the surface normal direction forming triangular prisms approximately characterised by identical volume (in blue). NCs (black dots) falling within the prisms are highlighted in red.

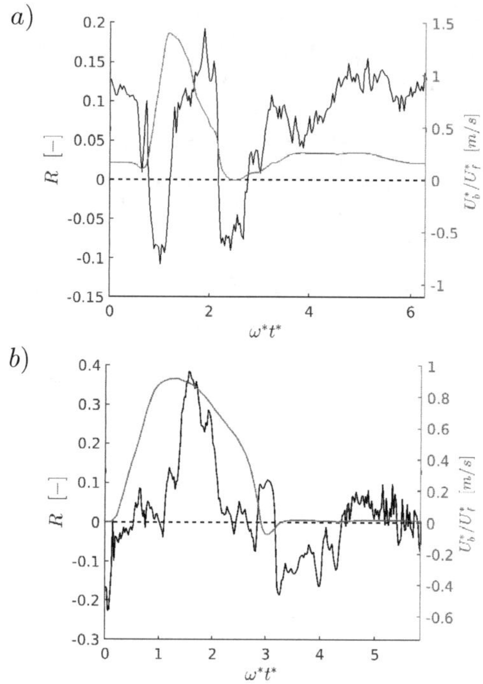

Figure 10. Time development of the cross-correlation R (black curve) computed for (a) run 1 and (b) run 2.

WSS is attained at the phases when bulk velocity is maximum (cf. figure 8b), in run 2 the maximum WSS is reached earlier (at $\omega^* t^* = 0.7$, cf. figure 8d). In fact, the origin of the peaks of WSS in run 1 is associated with the interaction of the vortical structures with the wall, the development of which is driven by the core flow. On the other hand, in run 2, the WSS is caused by viscous stresses that, like in a laminar oscillatory boundary layer, follow the maximum acceleration phase (at $\omega^* t^* = 0.2$) and forerun the maximum core velocity phase (at $\omega^* t^* = 1.2$). The magnitude of WSS in the mouse aortic arch is overall significantly higher than in the human case, both in dimensional and dimensionless terms, which is consistent with concepts of allometric scaling [8] and with data from literature [17]. In fact, in an oscillatory flow (assumed as a prototype), the peak WSS is proportional to $\varrho^* U_0^* \omega^* \delta^* = (\varrho^* \nu^*) Re_\delta \omega^*$, and the ratio $(Re_\delta \omega^*)_{mouse}/(Re_\delta \omega^*)_{human}$ is equal to 3.6. The maximum WSS in run 2 is approximately equal to 22 Pa, which corresponds to the value observed in vivo by Suo et al. [15]. Since the geometry is similar and the value of U_s^* is approximately the same for the two runs, also the contribution of the steady stream component to the WSS is larger in the mouse case.

3.1. Dispersion of NP in the near-wall region

In order to estimate the effectiveness of drug delivery or of diagnostic techniques based on the transport of NCs, particles are systematically released uniformly at the inlet cross-section and their trajectories are computed by numerically integrating the fluid velocity suitably interpolated at the particle position throughout the pulsation. The concentration of the particles at the inlet is irrelevant, since the passive tracers are released only to investigate the Lagrangian properties of the flow. Downstream of the inlet, NCs are dispersed by the flow and some of them eventually come in contact with the wall, thereby ending its motion. The capability of the flow to disperse the NCs uniformly close to the wall throughout the pulsation is evaluated by adopting the following procedure. A coarse uniform triangular tessellation of the wall surface is made (see figure 9) and tiles are slightly extruded in the radial direction to obtain regions of equal volume. The number of particles lying within individual regions is counted at each phase. Then, particle number concentration near the walls, n_p, is computed and related to the local WSS, τ_w^*, to identify regions where the WSS is larger than the value averaged over the wall surface and, simultaneously, n_p is smaller than the average value, which is an unfavourable combination to detect potential diseases at the vessel walls. The percentage of cases where such a combination is observed is 25% for run 1 and 28% for run 2. Nonetheless, during a pulsation, no regions are left unexplored by NCs. In order to determine the phases of the pulsation where fluid particles approach more frequently high-WSS regions, namely a condition that can be associated with a good diagnostic efficiency, the following definition of cross-correlation R is defined

$$R(t) = \frac{\left\langle n_p^{*\prime}(t)\tau_w^{*\prime}(t)\right\rangle_{surf}}{\sigma_{n_p}^*(t)\sigma_{\tau_w}^*(t)} , \tag{1}$$

where $n_p^{*\prime} = n_p^* - \langle n_p^* \rangle_{surf}$ and $\tau_w^{*\prime} = \tau_w^* - \langle \tau_w^* \rangle_{surf}$ denote the fluctuations of NC number concentration and WSS, respectively, $\langle\cdot\rangle_{surf}$ and σ indicating the average over the vessel surface and standard deviation of the respective quantities at time t. Figure 10 shows the value of R for the two runs. Large values of R, that can be interpreted as a fair diagnostic efficiency, are observed during the early decelerating phases of the pulsation, when the adverse pressure gradient promotes convective events enhancing the NC dispersion. The phases where R is negative indicate that the diagnostic efficiency of NCs lays below the average. Such phases, however can be observed only during the diastole, when the NC motion is modest, and, in the human case, during the early systole, when the favourable pressure gradient supports the flow re-laminarisation.

4. Conclusions

The properties of the blood flow in the prototypes of human and murine aortic arches were investigated by means of DNSs. In the murine aortic arch the flow was found to be laminar. In this case, the WSS was particularly intense around the inlet, where the uniform velocity distribution rapidly adapts to the boundary conditions, and in the zones downstream of high-curvature regions. The peak of WSS is reached earlier than the phases of maximum bulk velocity, namely when the oscillatory boundary layer provides the maximum contribution to the WSS. Instead, in the human case the separation of the boundary layer was observed during the systolic phases when the flow rate is maximum, which promoted the development of turbulent vortical structures downstream of high-curvature regions. The regions where the vortical structures interacted with the wall correspond to the zones of maximum WSS. Then, the vortical structures disappeared during the diastolic phases. The analysis of fluid-particle trajectories suggested that NCs potentially injected in the blood circulation would interact with the wall in any part of the aortic arch. Despite the flow in the human and murine aortic arches is remarkably different, the dispersion of NCs is good in both cases. In particular, the development of turbulent vortical structures in the human aortic arch was found to enhance the effectiveness of diagnostic techniques, which are based on the dispersion of NCs. In perspective, the present investigation will be extended to real three-dimensional models of the aortic arch obtained from the MRI of pathological patients in order to account for the geometric features that have been presently neglected and to provide a customised evaluation of the effectiveness of the diagnostic techniques.

5. Acknowledgments

The generous support provided by the organizing committee of the 18th OpenFOAM Workshop is also acknowledged.

References

[1] Atabek, H. B., & Chang, C. C. (1961). Oscillatory flow near the entry of a circular tube. Zeitschrift für angewandte Mathematik und Physik ZAMP, 12, 185-201.
[2] Atabek, H. B., Chang, C. C., & Fingerson, L. M. (1964). Measurement of laminar oscillatory flow in the inlet length of a circular tube. Physics in Medicine & Biology, 9(2), 219.
[3] Ayyaswamy, P. S., Muzykantov, V., Eckmann, D. M., & Radhakrishnan, R. (2013). Nanocarrier hydrodynamics and binding in targeted drug delivery: challenges in numerical modeling and experimental validation. Journal of nanotechnology in engineering and medicine, 4(1), 011001.
[4] Callaghan, F. M., & Grieve, S. M. (2018). Normal patterns of thoracic aortic wall shear stress measured using four-dimensional flow MRI in a large population. American Journal of Physiology-Heart and Circulatory Physiology, 315(5), H1174-H1181.
[5] Chen, W., Schilperoort, M., Cao, Y., Shi, J., Tabas, I., & Tao, W. (2022). Macrophage-targeted nanomedicine for the diagnosis and treatment of atherosclerosis. Nature Reviews Cardiology, 19(4), 228-249.
[6] De Negri Atanasio, G., Ferrari, P. F., Baião, A., Perego, P., Sarmento, B., Palombo, D., & Campardelli, R. (2022). Bevacizumab encapsulation into PLGA nanoparticles functionalized with immunouteroglobin-1 as an innovative delivery system for atherosclerosis. International Journal of Biological Macromolecules, 221, 1618-1630.
[7] De Negri Atanasio, G., Ferrari, P. F., Campardelli, R., Perego, P., & Palombo, D. (2021). Innovative nanotools for vascular drug delivery: the atherosclerosis case study. Journal of Materials Chemistry B, 9(41), 8558-8568.
[8] Greve, J. M., Les, A. S., Tang, B. T., Draney Blomme, M. T., Wilson, N. M., Dalman, R. L., ... & Taylor, C. A. (2006). Allometric scaling of wall shear stress from mice to humans: quantification using cine phase-contrast MRI and computational fluid dynamics. American Journal of Physiology-Heart and Circulatory Physiology, 291(4), H1700-H1708.
[9] Hopper, S. E., Cuomo, F., Ferruzzi, J., Burris, N. S., Roccabianca, S., Humphrey, J. D., & Figueroa, C. A. (2021). Comparative study of human and murine aortic biomechanics and hemodynamics in vascular aging. Frontiers in Physiology, 12, 746796.
[10] Hossaini Nasr, S., & Huang, X. (2021). Nanotechnology for targeted therapy of atherosclerosis. Frontiers in Pharmacology, 12, 755569.
[11] Naruse, T., & Tanishita, K. (1996). Large curvature effect on pulsatile entrance flow in a curved tube: model experiment simulating blood flow in an aortic arch.
[12] Saitta, S., Sturla, F., Caimi, A., Riva, A., Palumbo, M. C., Nano, G., ... & Redaelli, A. (2022). A deep learning-based and fully automated pipeline for thoracic aorta geometric analysis and planning for endovascular repair from computed tomography. Journal of digital imaging, 35(2), 226-239.
[13] Shahcheraghi, N., Dwyer, H. A., Cheer, A. Y., Barakat, A. I., & Rutaganira, T. (2002). Unsteady and three-dimensional simulation of blood flow in the human aortic arch. J. Biomech. Eng., 124(4), 378-387.
[14] Suito, H., Takizawa, K., Huynh, V. Q., Sze, D., & Ueda, T. (2014). FSI analysis of the blood flow and geometrical characteristics in the thoracic aorta. Computational Mechanics, 54, 1035-1045.
[15] Suo, J., Ferrara, D. E., Sorescu, D., Guldberg, R. E., Taylor, W. R., & Giddens, D. P. (2007). Hemodynamic shear stresses in mouse aortas: implications for atherogenesis. Arteriosclerosis, thrombosis, and vascular biology, 27(2), 346-351.
[16] Tuzi, R., & Blondeaux, P. (2008). Intermittent turbulence in a pulsating pipe flow. Journal of Fluid Mechanics, 599, 51-79.
[17] Weinberg, P. D., & Ethier, C. R. (2007). Twenty-fold difference in hemodynamic wall shear stress between murine and human aortas. Journal of biomechanics, 40(7), 1594-1598.
[18] Weller, H. G., Tabor, G., Jasak, H., & Fureby, C. (1998). A tensorial approach to computational continuum mechanics using object-oriented techniques. Computers in physics, 12(6), 620-631.
[19] Zhang, J., Zu, Y., Dhanasekara, C. S., Li, J., Wu, D., Fan, Z., & Wang, S. (2017). Detection and treatment of atherosclerosis using nanoparticles. Wiley Interdisciplinary Reviews: Nanomedicine and Nanobiotechnology, 9(1), e1412.

IOP Conf. Series: Materials Science and Engineering **1312** (2024) 012012 doi:10.1088/1757-899X/1312/1/012012

COMPUTATIONAL FLUID DYNAMICS STUDY ON THE INFLUENCE OF VARIABLE CANT ANGLE WINGLETS ON TOTAL DRAG REDUCTION

Giulia Innocenti[1], Eric Segalerba[1], Joel Guerrero[2]
[1]Dipartimento di Ingegneria Civile, Chimica ed Ambientale (DICCA), Università di Genova, Genova, Italy
[2]Wolf Dynamics SRL, Genova, Italy

May 29, 2024

Abstract

The development of winglets has been motivated by the need to reduce lift-induced drag in aircraft. One of the main limitations, since they are fixed devices, is related to their inability in adapting to variable flight conditions. To address this issue, this study proposes the use of variable cant angle winglets that can be adjusted to optimize drag reduction under different flight conditions. In this study, we use the CFD solver OpenFOAM to investigate the effects of varying the winglet cant angle. Instead of a fully mesh morphing approach, a quasi-steady approach was taken where different winglet configurations were investigated by changing the cant angle from 0° to 70°. The main reason for taking this approach was to reduce the burden related to the dynamic meshing approach. The results obtained in this flight condition show an optimal cant angle, with respect to the aerodynamic efficiency, around 40°.

1 Introduction

The main motivation behind the present study is the improvement and development of new wingtip devices technology for drag reduction. From an aerodynamic point of view, one of the main sources of losses in terms of drag is the lift-induced drag; it can amount to as much as 40% of the total drag at cruise conditions and 80–90% of the total drag during take-off and climb conditions[1]. The idea of wingtip fixed devices was developed in the late 1800s by British aerodynamicist F.W. Lancaster, at the beginning as a vertical plate designed for reducing drag by controlling wingtip vortices [2], which are generated by the pressure imbalance among pressure and suction sides; those are released at the trailing edge and produce a downward flow near the wing, which interacts with the free-stream flow and causes a drag increase. Whitcomb, an engineer at NASA Langley Research Center, was inspired by the observation of nature, in particular, by the soaring birds and their use of tip feathers to control flight. He mastered the winglet design, to improve aircraft performance [3, 4].

2 Wing model

The wing used in this study was taken from reference [5] and the CAD can be retrieved at the following link [6]. The cross-section is based on the NACA 65210, which is known for its high lift and low drag characteristics. The maximum chord (C_{max}) of the wing is 0.985 $[m]$. The ratio between the root chord and the tip chord is equal to 3.58 and the aspect ratio AR is 7.26. The geometry of the wing has been modified to create eight different configurations by changing the winglet cant angle θ from 0° to 70°; Figure 1 shows the wing geometry with a cant angle equal to 30°.

The mesh was created using the mesh generation utility, snappyHexMesh, distributed with OpenFOAM. In Figure 2, we depict the mesh details close to the wing surface and the inflation layer. The mesh is made up of around 5 million cells with an acceptable overall quality: maximum non orthogonality value equal to 64 degrees, and a skewness less than 4.

IOP Conf. Series: Materials Science and Engineering **1312** (2024) 012012 doi:10.1088/1757-899X/1312/1/012012

Figure 1: Wing geometry. In the figure, the winglet cant angle is equal to $\theta = 30°$

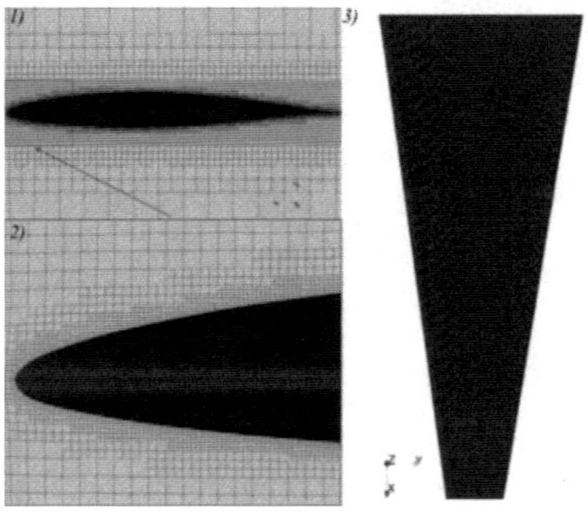

Figure 2: Wing mesh.

The computational domain dimensions are 4xC_{max} along the x-direction, 4xC_{max} in front of the wing and 12xC_{max} behind the wing in the y-direction, and 8xC_{max} in the z-direction. This dimensions ensure a small blockage ratio (lower than 0.8%) and avoids numerical errors related to large gradients close to the boundaries. The streamwise length allows for a complete development of the flow in the wake. The angle between the airfoil chord line and the direction of the undisturbed free-stream V_∞, the angle-of-attack AOA, is fixed equal to 0°. In Figure 3 sketches the name of the patches, whose boundary conditions are:

- inlet, uniform velocity field equal to 71.5 m/s, corresponding to a Reynolds number of 3 million (based on C_{max});
- outlet, pressure condition equal to 0 $[Pa]$;
- top, bottom, left and right, slip wall condition;
- wing, no-slip condition.

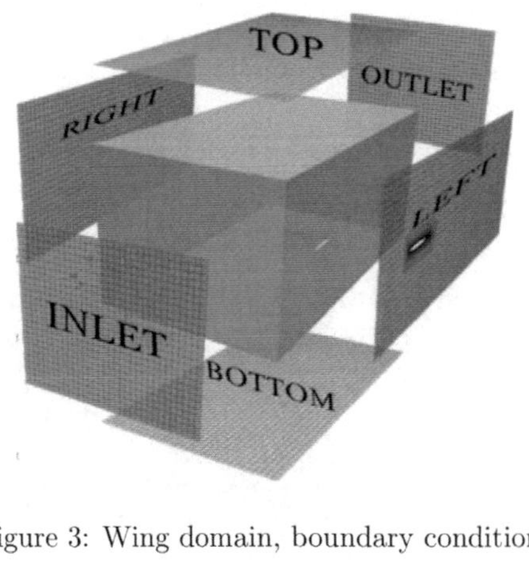

Figure 3: Wing domain, boundary conditions

IOP Conf. Series: Materials Science and Engineering **1312** (2024) 012012 doi:10.1088/1757-899X/1312/1/012012

3 Numerical method, results and validation

The incompressible Reynolds-Averaged Navier-Stokes (RANS) equations are numerically approximated by using the open-source finite volume solver OpenFOAM [13]. The cell-centered values of the variables are interpolated at the face locations using a second-order central difference scheme for the diffusive terms. The convective terms are discretized by the so-called limited linear scheme, a second-order accurate bounded total variation diminishing (TVD) scheme, resulting from applying the Sweby limiter to the central differencing scheme to enforce monotonicity [14]. The pressure-velocity coupling is achieved through the SIMPLE algorithm [15]. For the turbulence modeling, the $k-\omega$ turbulence model is used [7, 8].

Figure 4 shows contours of pressure for θ equal to 0° and 70°. Pressure values are higher at the leading and trailing edges where the flow velocity is lower, both for θ equal 0° and 70°. For increasing values of the cant angle, the range of pressure (minimum and maximum) increases due to a general decrease of the flow velocity on the wing related to the presence of the winglet which forces a recirculation zone close to the point in which the wing slope changes.

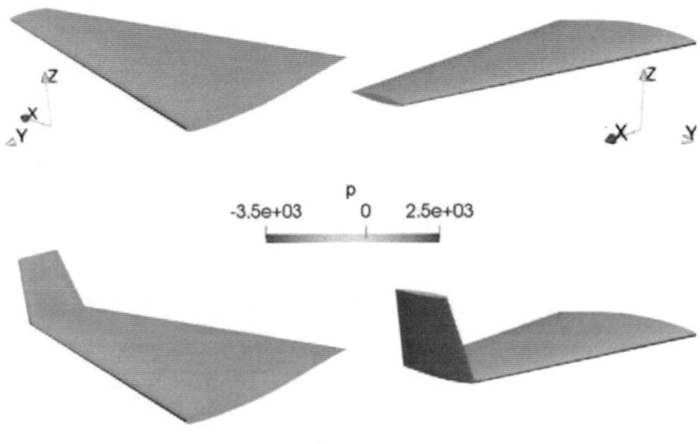

Figure 4: Pressure contours, winglet cant angle $\theta = 0°$ and $\theta = 70°$

In Figure 5 we depict the distribution of y^+ on the wing for the same extreme angles. The average y^+ value is equal to 54 meaning that the first cell is placed in the log-law layer and corresponds to a wall modeling simulation.

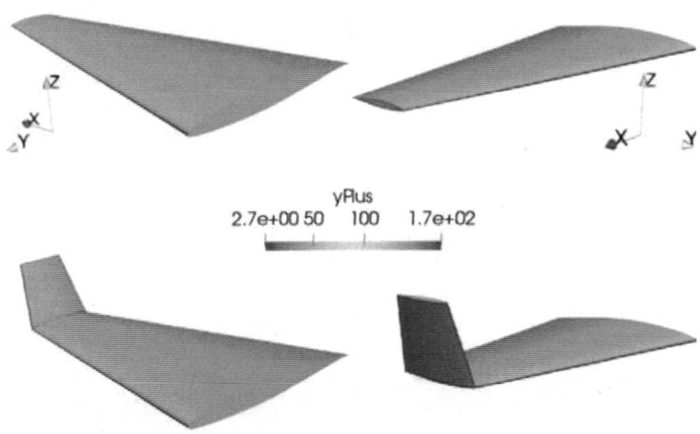

Figure 5: Dimensionless wall distance y^+ contours, winglet cant angle $\theta = 0°$ and $\theta = 70°$

Aerodynamic coefficients are non-dimensional parameters that represent the lift, drag, and pitching moment. In order to compare the present results, it is fundamental to define and analyze the trend of these parameters; their non-dimensionality allows the comparison between different aerodynamic

IOP Conf. Series: Materials Science and Engineering **1312** (2024) 012012 doi:10.1088/1757-899X/1312/1/012012

bodies characterized by different sizes, shapes, and orientations.

$$c_d = \frac{D}{\frac{1}{2}\rho\, V_\infty^2\, S} \qquad c_l = \frac{L}{\frac{1}{2}\rho\, V_\infty^2\, S} \qquad c_m = \frac{M}{\frac{1}{2}\rho\, V_\infty^2\, S\, c}$$

where D is the total drag force, L is the lift force, ρ is the air density, V_∞ the free-stream velocity, S is the wing reference area and c is the length of the mean aerodynamic chord.
In Figure 6, the lift coefficient is shown: it exhibits an increasing trend which reaches its best performances with $\theta = 50°$; when compared to the plain wing, the configuration with the winglet with that cant angle, produces an increase of 2 % in terms of lift. Considering a wing of NACA 65210 airfoil section [5], the lift coefficient is equal to 0.1150 (AOA = 0°) and in the present study the lift coefficient for the plain wing is equal to 0.1164; the two values are very closed and this provides a validation of the results.

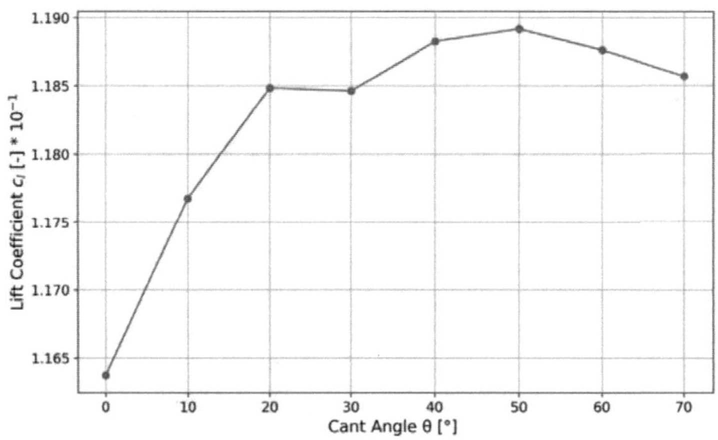

Figure 6: Lift coefficient c_l versus the winglet cant angle θ

In Figure 7, the drag coefficient increases by modifying the cant angle and it reaches the maximum value for $\theta = 70°$. Around $\theta = 30°$, there is a slight reduction in the drag reduction and then it starts to increase again monotonically.

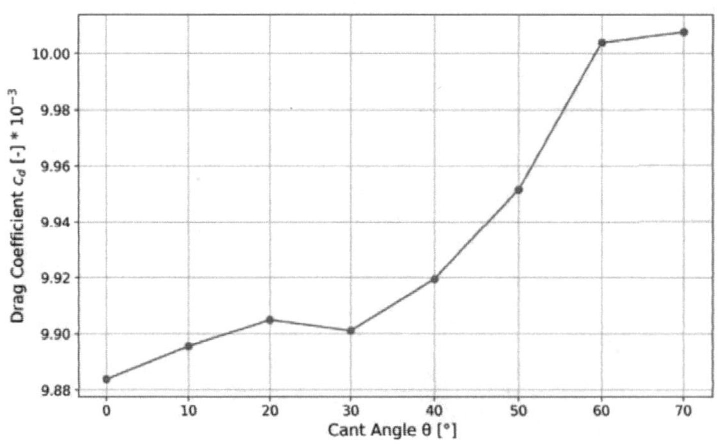

Figure 7: Drag coefficient c_d versus the winglet cant angle θ

The following equation defines the total drag coefficient c_d:

$$c_d = c_{d0} + c_{di} \qquad c_{di} = \frac{c_l^2}{\pi\, AR\, e}$$

where c_{d0} is the base drag coefficient at zero lift, c_{di} is the induced drag coefficient, AR is the aspect ratio and e is the efficiency factor.
To estimate the induced drag coefficient c_{di}, the efficiency factor e has been estimated from the reference paper [5]. The efficiency parameter is calculated for the condition of an angle of attack equal to 0°, the aspect ratio AR is equal to 9, and the values of c_d, c_l and c_{d0} have been extracted from the aerodynamic coefficients plots. Its value is equal to 0.71, reasonable because e is between 0 and 1.
The induced drag coefficient for the present case is evaluated by fixing the denominator of the c_{di} as a constant parameter equal to the reference case, to scale the problem in the same way.

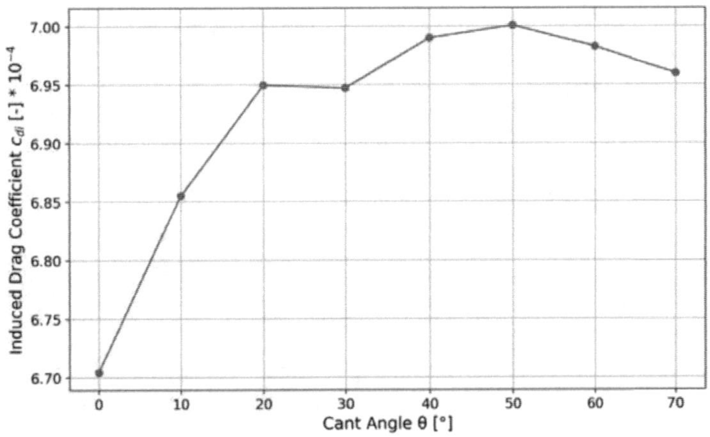

Figure 8: Induced drag coefficient c_{di} versus the winglet cant angle θ

Aerodynamic efficiency is defined using the ratio of a body's lift to drag, equivalent to the ratio between the lift and drag coefficients. An efficient design maximizes feasible lift while minimizing drag. In Figure 9, the aerodynamic efficiency is plotted against the cant angle, and the best position of the winglet, to maximize aerodynamic efficiency, is obtained for $\theta = 40°$. It is important to underline that this is the optimal cant angle for the specific flight condition and it provides an improvement of this parameter by almost 2% compared to the plain wing.

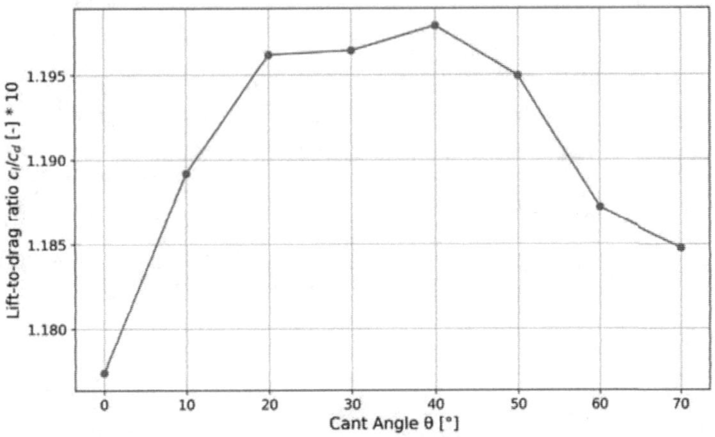

Figure 9: Lift-to-drag ratio c_l/c_d versus the winglet cant angle θ

Pitching moment coefficient c_m is a fundamental parameter for the static stability of aircraft.
In Figure 10 this quantity is plotted and it can be noticed that it presents a monotonically decreasing behavior with the cant angle. Interestingly, the winglet positions do not largely influence the pitching moment coefficient, as the values are very similar; this is a good point to take into account considering

all the benefits of these devices. The pitching moment coefficient is equal to 0.0359 (AOA = 0°) for the reference wing [5] and in the present study for the plain wing it is equal to 0.0367.

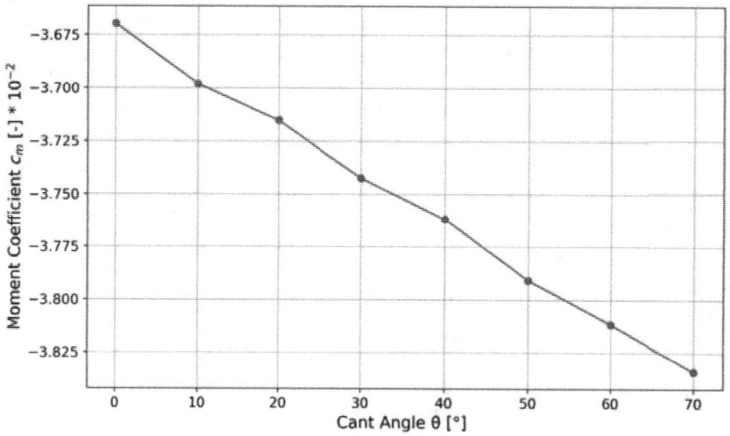

Figure 10: Pitching coefficient c_l versus the winglet cant angle θ

4 Conclusions and future perspectives

In the present contribution, a preliminary work on how different cant angles of winglets can affect the aerodynamical coefficient is presented. The steady simulations for eight cant angles are performed for the same flight condition, with an angle of attack equal to 0°. Winglets increased maximum lift and, furthermore, the wing performance, in terms of the aerodynamic efficiency, improves, in particular for a cant angle of 40°, which maximizes the lift minimizing the drag. On the other side, the induced drag tends to increase with θ. It would be necessary to conduct a global optimization study to find the best configuration for this flight condition.

These preliminary results provide a solid basis for more advanced analysis which will take into account also the dynamic effects of a real-time change in winglet's position, which might allow for a real-time optimization of the geometry to account for the different flight conditions an aircraft encounters in the flight. Moreover, a dynamic approach would also consider an effect similar to the dynamic stall experienced by helicopters' rotor whose angle of attack changes dynamically. For this purpose, a model to control the rotation of the wing around arbitrary axes and positions is developed. These results will provide a time-continuous measurement of the aerodynamical coefficients. In this future contribution, it will be taken into account also the effect of different flight conditions.

5 Aknowledgement

The authors acknowledge the organizing committee of the 18th OpenFOAM Workshop for the support provided.

6 Data availability

CAD model can be downloaded from the following link [6].
The sample case can be downloaded from the link [9].

References

[1] J.E. Guerrero, M. Sanguineti, and K. Wittkowski. Variable cant angle winglets for improvement of aircraft flight performance. *Springer: Meccanica*, 55:1917–1947, 2020.

[2] J.E. Guerrero, D. Maestro, and A. Bottaro. Biomimetic spiroid winglets for lift and drag control. *Elsevier: C.R. Mecanique*, 340:67–80, 2012.

[3] J.R. Chambers. Concept to reality: Contributions of the langley research center to u.s. civil aircraft of the 1990s. *NASA History Series SP-2003-45299; DIANE Publishing Company: Collingdale, PA, USA*, 2003.

[4] R.T. Whitcomb. A design approach and selected wind tunnel results at high subsonic speeds for wing-tip mounted winglets. *Technical Report NASA-TN-D-8260; NASA Langley Research Center: Hampton, VA, USA.*, 1976.

[5] James C. Sivells. Experimental and calculated characteristics of three wings of naca 64-210 and 65-210 airfoil sections with and without 2° washout. *National advisory committee for aeronautic*, 1947.

[6] https://cad.onshape.com/documents/23badb97765070e18fa25287/w/0d277e8bb725a19c7bebcc0f/e/c1dfe9d9c41b632b5aeb19c1?rendermode=0uistate=662fa4ea67acc6051ad64cc7. .

[7] David C. Wilcox. Turbulence modeling for cfd. *(2nd ed.)*, 1998.

[8] David C. Wilcox. Formulation of the k–ω turbulence model revisited. *AIAA Journal*, 46 (11): 2823–2838, 2008.

[9] https://github.com/ericsegalerba/openfoam_proceedings.git. .

IOP Conf. Series: Materials Science and Engineering **1312** (2024) 012013 doi:10.1088/1757-899X/1312/1/012013

Open source tools for OpenFOAM - Adaptive mesh refinement and convergence detection

Wouter Remmerie[1] and Nikola Majksner[2]

[1]CEO, AirShaper, Antwerp, Belgium
[2]Research and Development, AirShaper, Antwerp, Belgium

E-mail: wouter@airshaper.com

Abstract. This paper presents two open-source tools to increase precision and automate CFD simulations in OpenFOAM. The first is an adaptive mesh refinement method which uses custom field properties to refine both volume and surface cells. The second is a method to detect convergence and dynamically size the length of the averaging window to obtain reliable averaged values.

1 Adaptive mesh refinement

1.1 Problem statement

The adaptive mesh refinement library builds on snappyHexMesh, the built-in mesh generator of OpenFOAM. By default, snappyHexMesh works as follows:

- Create a castellated mesh: a hexagonal mesh which can be refined based on user input / geometrical features (like refinement boxes to capture the wake, refinement around sharp edges, ...).
- Snapped mesh: then, that mesh is cut by the actual geometry (the black circle in Fig 1 and Fig. 2). The cut part of the mesh is then "snapped" to the surface of the geometry to match the actual shape.
- Optionally, layers can be added.

When refining a mesh after the snapping phase (to refine the wake for example), the surface quality will not be improved – see Fig. 1. OpenFOAM does provide AMR (adaptive mesh refinement) functionality through the 'dynamicRefineFvMesh' library ([1]). However, this functionality does not revert to the original surface geometry and only refines and unrefines existing cells. It is therefore highly useful to dynamically refine transient flow patterns (multiphase flows like the dam break tutorial or vortex shedding) but not suitable to refine the surface mesh on steady state (or transient) flows. The desired behaviour is shown in Fig. 2, where the cells are refined and then cut again using the original geometry.

1.2 Suggested solution

To solve this, a method was implemented that stores the original castellated mesh, performs the refinement and snaps the refined mesh to the real surface. This process is structured as follows (see also Fig. 3):

IOP Conf. Series: Materials Science and Engineering **1312** (2024) 012013 doi:10.1088/1757-899X/1312/1/012013

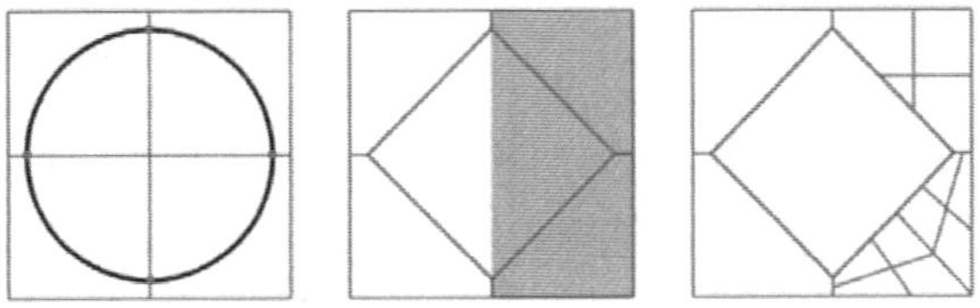

Figure 1: Refining cells after the snapping phase

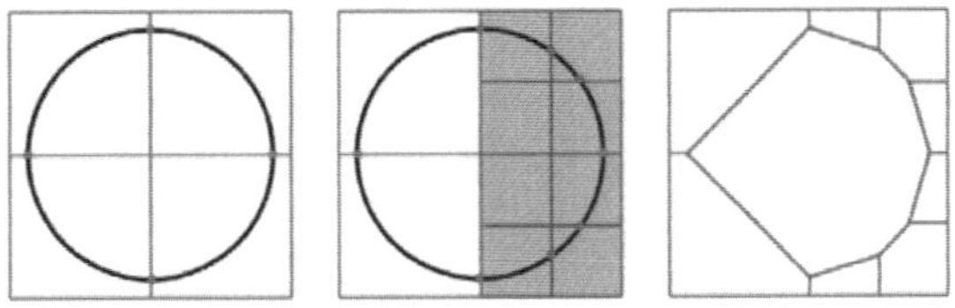

Figure 2: Refining cells before the snapping phase

1.2.1 Original mesh

- Castellated mesh: the original castellated mesh is created. Before the cutting starts, this castellated mesh is stored.
- Snapped mesh: the mesh is then snapped, all according to the standard snappyHexMesh procedure.
- CFD simulation: the CFD simulation is then performed and custom fields like the gradient of pressure and the vorticity are calculated. These are then mapped onto the original castellated mesh.

1.2.2 Refined mesh

- Cells with values above a custom threshold will be refined (using the original castellated mesh) by pimpleFoam (not to run CFD).
- This refined mesh is then cut by the geometry, making the newly refined cells adhere to the surface (using snappyHexMesh).
- Finally, a CFD simulation is performed on the refined mesh.

1.3 Code description

The repository [2] provides adaptive mesh refinement for both the volume mesh and surface mesh. Any field can be used for refinement. In the tutorial, two options have been included (for OpenFOAM v2012, but also compatible with v2312):

- grad(p): the pressure gradient
- curl(U): the curl of the velocity field

Both are multiplied by the cell length to "discourage" the algorithm from refining the smallest cells. The method has 3 major parts:

IOP Conf. Series: Materials Science and Engineering **1312** (2024) 012013 doi:10.1088/1757-899X/1312/1/012013

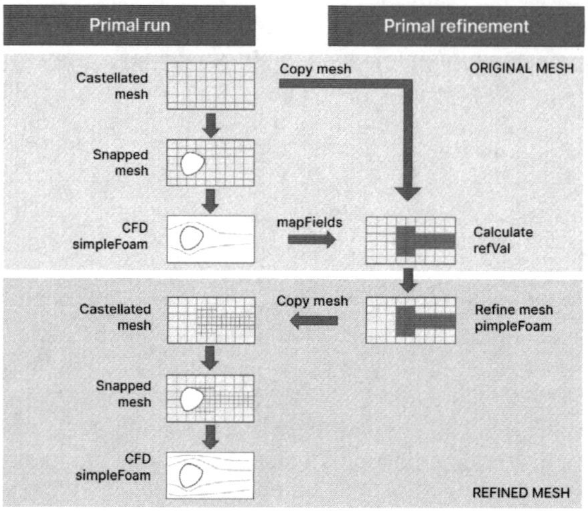

Figure 3: Process flow for adaptive mesh refinement

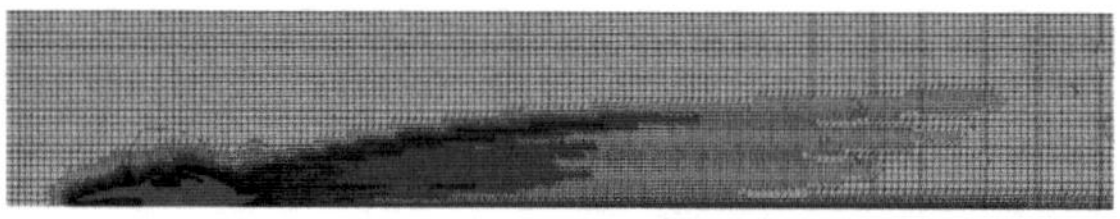

Figure 4: Wake refinement based on the curl of the velocity field

1.3.1 Initial mesh

SnappyHexMesh is run in two separate steps:

- castellated mesh: only runs the refinement phase using "baffles" for the surface of the geometry (to avoid the mesh inside the geometry being cut and to keep all cells hexagonal). This mesh is stored in primal_refinement for use in the mesh refinement phase.
- snapping phase: this part will cut the mesh (by also having castellated enabled) and snap it to the surface (by setting the surface patch type to "wall"). This mesh is used in primal_run for the first CFD loop.

1.3.2 Refinement field
Once the first CFD loop has finished, it maps the fields onto the stored castellated mesh (in primal_refinement) and calculates the refinement field (refVal), for which the lower limit value should be set manually in dynamicMeshDict.

1.3.3 Mesh refinement
The refinement is then applied to the castellated mesh of the first loop. Once refinement is done, the castellated mesh is copied to primal_run where the snapping phase is executed again to snap the refined mesh to the surface. This mesh is then used for the second CFD run.

1.4 Examples
Fig 4 illustrates the refinement of the wake by using the curl of the velocity as a refinement parameter. In this case, the method was applied multiple times, leading to a wake refinement featuring multiple refinement levels. It can be seen that the refinement is most intense close to the car (in this case, a 3D model of a Formula One car).

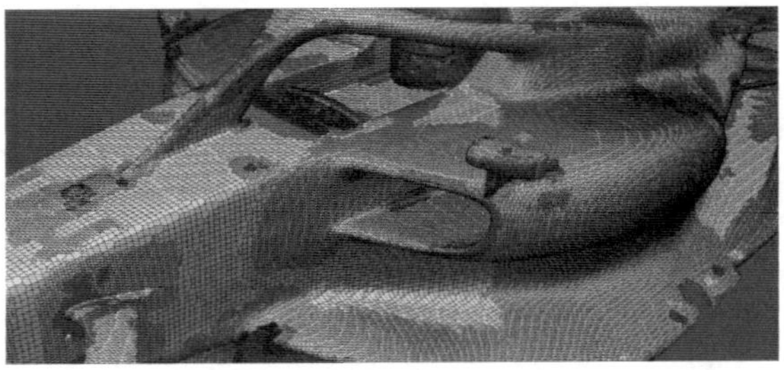

Figure 5: Surface refinement based on the gradient of the pressure field

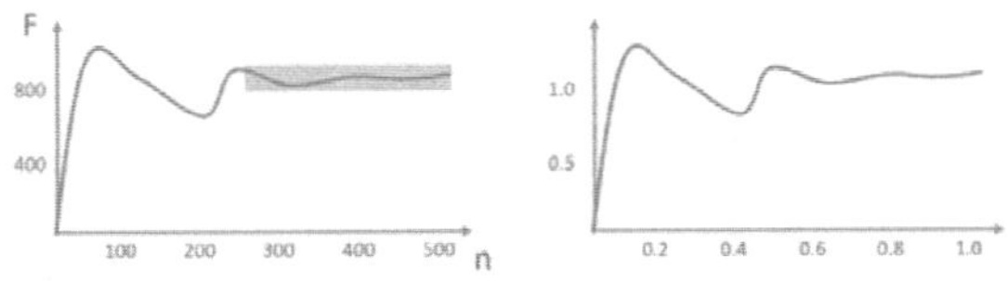

Figure 6: Normalization

Fig. 5 illustrates the refinement on the surface by using the gradient of the pressure field as a refinement parameter. This triggered refinement of the mesh in high-pressure gradient areas such as the flow around the suspension arms, the air intake at the side pods, the vertical beam of the halo etc.

2 Convergence detection

2.1 Problem statement

In addition to monitoring residuals, a common method to assess whether a simulation has converged is to look at the forces on an object. Especially for external aerodynamics, the drag, lift and lateral forces are usually of great interest (towards drag and lift coefficient, efficiency and power estimates, ...). It therefore makes sense to monitor these properties and wait until they are stable before starting the averaging.

OpenFOAM [3] already provides such functionality through run-time control function objects [4]. These allow, for example, to terminate the simulation as soon as the averaged drag coefficient is within a certain specified tolerance for a given number of steps. However, these parameters need to be determined manually for each case, making it difficult to use towards automation.

The open-source library presented here has been developed to provide fully automated convergence detection and averaging of OpenFOAM output data but is generally applicable to any data output for which convergence needs to be detected. It's been designed to automatically scale to the amplitude and length of the provided data, eliminating user input.

2.2 Suggested solution

2.2.1 Convergence detection

The theory behind the library is to mimic "human behaviour" to assess whether a curve has converged or not, following these steps:

- Normalization When we evaluate a curve as a human, we do so looking at its proportions: how much does it oscillate versus the total height, for how long has the curve been flat versus the total length of the curve, etc. So we look at relative properties, which can mathematically be represented by normalizing the curve on both the horizontal axis (the number of iterations n) and the vertical axis (the force in this case, but this could be any parameter). In this library, we use the average force value of the last 50 percent of iterations (a window size of 0.5) to normalize the entire curve (and we do this again for each iteration). Fig. 6 illustrates this.
- Gradient calculation Once we have normalized the curve, we want to assess when the curve has "stabilized. That means that the last bit of the curve somehow needs to have been "stable

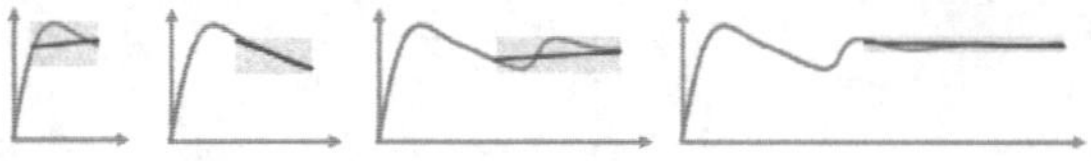

Figure 7: Curve fitting and gradient calculation

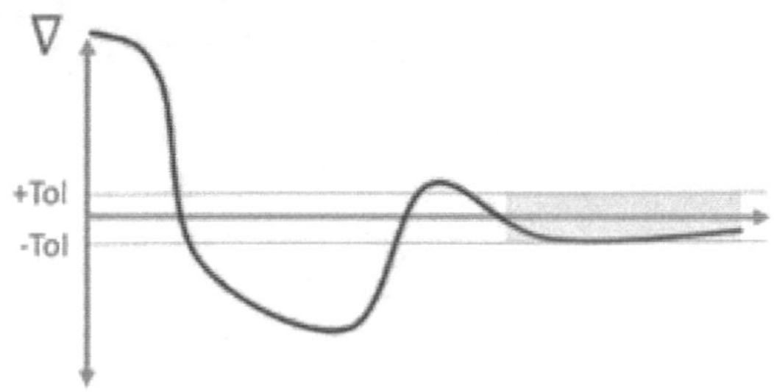

Figure 8: Threshold analysis

enough" for "long enough". For each iteration, we fit a straight line through the last 50 percent (a window size of 0.5) of the normalized curve and store the gradient of this curve. Fig. 7 illustrates this.

- Threshold This "running gradient" then indicates how stable the curve is (looking backwards) for each iteration. As we are working with a normalized curve, we can then define an absolute threshold value (irrespective of the magnitude of the force or the number of iterations) for this gradient to be within before we label a simulation as converged. We have chosen to evaluate this running gradient for the last third (so a window size of 0.33) of the curve. Fig. 8 illustrates this.

2.2.2 Averaging

Dynamic window sizing - Once convergence has been detected, the simulation will start to average. A modified version of the conversion detection method will detect when the averaging has taken long enough to provide a reliable averaged value / reliable averaged fields – see Fig. 9 (the yellow bar indicates the convergence detection, the green one the averaging window).
This detection method differs from the convergence detection method in the following ways:

- Normalization is frozen and the horizontal axis index will grow beyond 1. The longer a simulation runs, the lower the gradient value will become (as you divide by an ever-increasing index).
- The threshold values for the gradient are more strict (+-0.025 versus +-0.075).
- The evaluation window is now the entire length since the convergence point (versus a 0.5 window for the convergence detection), as there is no need any more to ignore initial oscillations. The whole averaging window needs to feature a stable curve.

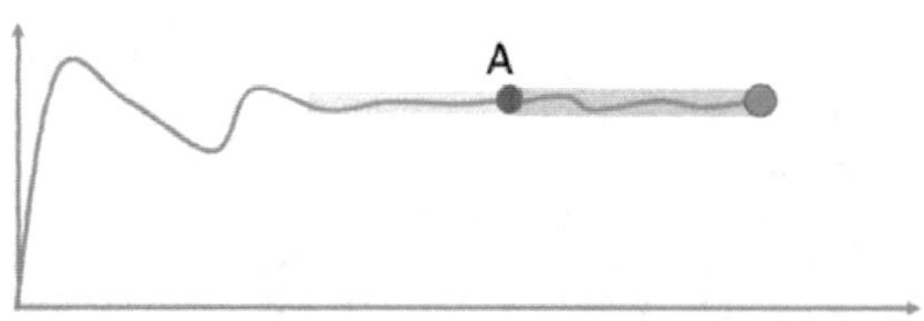

Figure 9: Dynamic sizing

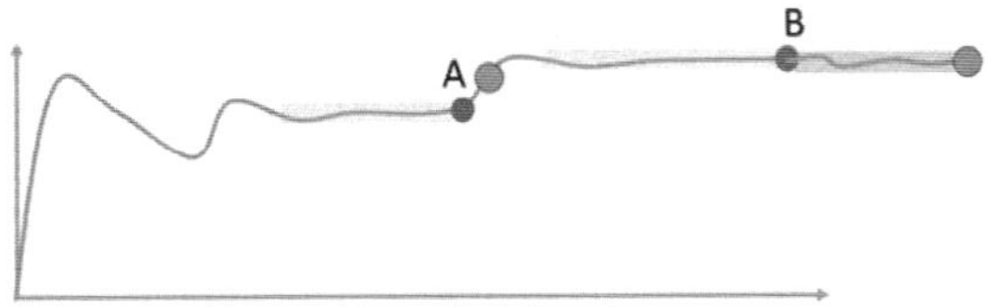

Figure 10: Re-start of convergence detection

Re-start - In the background, this convergence detection method keeps running: sometimes, after convergence has been detected, the curve can go up or down and the gradient can "exit" the threshold band again. Fig. 10 illustrates how at point A the curve seemed to have converged (looking back at a flat part) but then suddenly rose (see the orange dot). In that case, the state of the simulation will again be set to "not converged" until the convergence criterion is met again (at point B in the curve below). That also means averaging will be stopped until convergence is reached again.

2.3 Code description
The principles laid out above have been implemented in the repository [5]. The main parameters are described below (for OpenFOAM v2112, but also compatible with v2312):

- windowNormalization - default value 0.5: This parameter determines the length of the window used for normalizing the curve. Default value is 0.5, meaning that the entire curve will be divided by the averaged value of the last 0.5 (or 50 percent) of the curve.
- windowGradient - default value 0.5: This parameter determines the length of the window used to calculate the gradient. The default value of 0.5 means that the gradient will be fitted through the last 0.5 (or 50 percent) of the curve.
- windowGradientAveraging - default value 0.99: Same as above, but for the averaging phase (once convergence has been detected). Nearly the entire window is used here, since convergence has already been detected and the curve is in a "steady" phase (i.e. the initial oscillations are gone).
- windowEvaluation - default value 0.333: This parameter determines the length of the window used to evaluate the running gradient. The default value of 0.333 means the algorithm will evaluate the last third of the curve (to see if the gradient was within the specified threshold within that window).
- windowEvaluationAveraging - default value 0.333: Same as above, but for the averaging phase (once convergence has been detected).
- thresholdConvergence - default value 0.075: This is the threshold value within which the gradient needs to be during the evaluation window. The default value is 0.075 for convergence detection, meaning the gradient needs to remain within [-0.075 0.075] before convergence is detected.
- thresholdAveraging - default value 0.025: Same as above but for the averaging phase. The default value of 0.025 is more stringent compared to the one for convergence, to get reliable averaged values and fields.
- iterationMaxConvergence - default value 4000: This is a hard stop to detect convergence. The default value is 4000 iterations to force averaging.
- iterationMaxAveraging - default value 2000: The averaging will be stopped after 2000 iterations, whether it detected completion or not. So in total, a simulation will never run for more than 4000+2000=6000 iterations.
- iterationMinConvergence - default value 200: A simulation will run for at least 200 iterations before convergence can be detected.
- iterationMinAveraging - default value 200: The averaging window will be at least 200 iterations long.

IOP Conf. Series: Materials Science and Engineering **1312** (2024) 012013 doi:10.1088/1757-899X/1312/1/012013

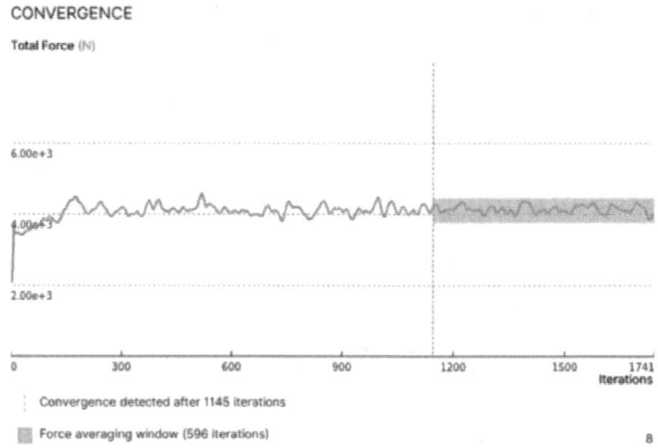

Figure 11: Convergence detection and averaging window sizing example on a real industrial case.

- forceStabilityFactor - default value 2: During averaging, there is a check to see if the forces go outside the [0.5*average ; 2*average] range. If this happens, it indicates a run-away of forces and the simulation is likely going to crash.

2.4 Examples
Fig. 11 illustrates the convergence detection and dynamic sizing of the averaging window for a real industrial case. Mind this graph represents the "second" CFD run, on the adaptively refined mesh. It started from the mapped fields from the first, coarse mesh and therefore the initial oscillations are fairly limited. This illustrates the method works well even on a fairly irregular force graph, with a varied "frequency" content.

3 Conclusion
The presented open-source libraries for OpenFOAM allow users to optimize their workflow by automatically and adaptively refining the mesh and detecting convergence / sizing the length of the averaging window. These methods have been applied successfully in thousands of industrial cases. The code is freely available to the community and has already been evaluated/implemented by OEMs and experts within the OpenFOAM community.
However, there is still room for improvement. The adaptive mesh refinement process can be quite time-consuming. Perhaps a modified version looking at the frequency spectrum could be a good upgrade. The convergence detection library features quite a high number of arbitrary control parameters, so perhaps some of these could be eliminated, or perhaps a version using the frequency content of the graph could be implemented.

4 Acknowledgements
Both repositories have been made publicly available [6] and have been presented at the Aero Vehicles conference in Poitiers [7] and the 18th OpenFOAM workshop in Genova [8]. The authors would like to thank these organizations for their support.

References

[1] https://openfoamwiki.net/index.php/Parameter_Definitions_-_dynamicRefineFvMesh.

[2] AirShaper adaptive mesh refinement code repository. https://github.com/airshaper/adaptive-mesh-refinement. Accessed on Fri, May 10, 2024.

[3] OpenFOAM — Free CFD Software — The OpenFOAM Foundation. https://openfoam.org/. Accessed on Mon, May 13, 2024.

[4] run-time control function objects. https://www.openfoam.com/documentation/guides/latest/doc/guide-fos-utilities-runtime-control.html. Accessed on Tue, January 02, 2024.

IOP Conf. Series: Materials Science and Engineering **1312** (2024) 012013 doi:10.1088/1757-899X/1312/1/012013

[5] AirShaper convergence detection code repository. https://github.com/airshaper/openfoam-convergence-detection. Accessed on Fri, May 10, 2024.

[6] AirShaper Repositories. https://github.com/airshaper. Accessed on Fri, May 10, 2024.

[7] Aerovehicles 5 - Sciencesconf.org. https://aerovehicles5.sciencesconf.org/. Accessed on Fri, May 10, 2024.

[8] 18th OpenFOAM Workshop 2023. https://openfoam-extend.sourceforge.net/OpenFOAM_Workshops/OFW18_2023_Genoa/index.html. Accessed on Fri, May 10, 2024.

AUTHOR INDEX